Resilient Contractor Safety Management

Contractor management has become an essential component of operational success, risk mitigation, and regulatory compliance across industries. As hiring clients place increasing demands on their contractors, understanding the expectations and the reasons behind them in the context of safety is more important than ever. This book offers insights into the evolving landscape of contractor engagement.

This book, which is the first of its kind, delivers a structured and practical guide covering every stage of contractor safety management, from selection, prequalification, and onboarding, to assessments, training, incident prevention and investigation, and much more. Each chapter presents practical guidance based on real-world applications, with attention given to the growing role of AI and future trends. The key benefit for the reader is a better understanding of how to meet client requirements with confidence, while strengthening long-term professional relationships and ensuring that safety is at the forefront of the relationship.

Resilient Contractor Safety Management is designed for contractors who require guidance on safety management. This book can also help hiring clients with valuable insights into how to establish and nurture productive relationships with their contractors. Additionally, it will appeal to subcontractors, project managers, procurement professionals, compliance officers, and anyone involved in contractor oversight.

Resilient Contractor Safety Management

Edited by
Simon Goncharenko

CRC Press
Taylor & Francis Group
Boca Raton London New York

CRC Press is an imprint of the
Taylor & Francis Group, an **informa** business

Designed cover image: iStock. Akarawut Lohacharoenvanich

First edition published 2026
by CRC Press
2385 NW Executive Center Drive, Suite 320, Boca Raton FL 33431

and by CRC Press
4 Park Square, Milton Park, Abingdon, Oxon, OX14 4RN

CRC Press is an imprint of Taylor & Francis Group, LLC

Library of Congress Cataloging-in-Publication Data
Names: Goncharenko, Simon (Industrial safety specialist) editor
Title: Resilient contractor safety management / edited by Simon Goncharenko.
Description: First edition. | Boca Raton : CRC Press, 2026. |
Includes bibliographical references and index.
Identifiers: LCCN 2025051930 | ISBN 9781041151081 hbk | ISBN 9781041154419 pbk |
ISBN 9781003679448 ebk
Subjects: LCSH: Industrial safety—Management | Contracting out—Management
Classification: LCC T55 .R46 2026
LC record available at https://lccn.loc.gov/2025051930

ISBN: 978-1-041-15108-1 (hbk)
ISBN: 978-1-041-15441-9 (pbk)
ISBN: 978-1-003-67944-8 (ebk)

DOI: 10.1201/9781003679448

Typeset in Times
by Apex CoVantage, LLC

To my family – for reminding me that safety is always about people, not numbers, and to all the frontline workers whose lives will be saved by the lessons of this book.

Contents

About the Editor

Dr. Simon Goncharenko (PhD, CSP) leads Expert Services at Veriforce, teaches Policies and Ethics of Health and Safety at Liberty University, USA, and is reshaping safety culture worldwide via his Save Lives Global initiatives, originated following the publication of Save Lives: Pushing Boundaries in Human Factors (SaveLives-Global.com). With over 30 years of field and academic experience across three continents, Dr. Goncharenko is a respected voice in operational excellence and human factors. He has worked with global giants, including Meta, CBRE, and BHP, has authored or contributed to over 80 articles and six books, and developed the world's first human factors training program for mission-critical facilities.

Contributors

Karl DeLooff
Leadership and Safety Consultant with SB2 Services
USA

Chris Detillier
Senior HSE Advisor – Veriforce LLC
USA

Paul Espenan
Sr Vice President EHS&R – Diversified Energy Company
USA

Chris Gafford
EHS Manager – Ring Energy
USA

Ron Gantt
Vice President EHS – Beale Infrastructure
USA

Rick L Ingram
Vice President – GOARC
USA

Thomas N. Meek, Jr.
Vice President of Compliance Operations – Veriforce LLC
USA

Josh Ortega
Vice President of Global Health, Safety and Sustainability – Veriforce LLC
USA

Samuel Rayborn
Director of Purchasing Enablement and Field Support – QXO
USA

Jim Wetherbee
Captain, US Navy (Retired) and Former Astronaut – NASA
USA

Preface

I was sitting in a large conference room in The Woodlands, TX, a suburb of Houston, and as the speaker talked about the importance of contractor safety management, it suddenly dawned on me that there is a gap in this area. Sure, there are a lot of self-proclaimed experts, a few web-based solutions, and a plethora of short presentations, webinars, opinion pieces, and possibly even some white papers. But no book.

A quick internet search ensued, where I tested my hunch. I pulled up a few book titles that sounded like they were going to cover contractor management subject, but a closer evaluation of every single one of these resources revealed that they all fell short of the comprehensive promises their titles made. Or they chose to focus on just one industry. Or they zoomed into a specific aspect of the question with no consideration for a wider angle.

The problem was identified. Now, the question of solving the problem remained. While I may not possess the comprehensive knowledge of all the major aspects of managing contractors, I knew enough people who happened to be experts in many individual aspects of this subject. So, if a book were to be created, it would have to be a multi-author conglomeration instead of a single-author piece. The next step was to test the interest of the individual players who would have to be brought to the table.

One by one, I started reaching out to these individuals and was pleasantly surprised by the overall response. The vast majority of the potential contributors contacted were as excited as I was about the project and eager to support it and jump on board. Of course, as with any new thing, there were a few individuals who were simply not as convinced of the prospects of this project and gracefully bowed out. At the end, however, just the right number of experts agreed to participate.

When I say just the right number of individuals agreed to participate, I mean that we ended up with enough experts to cover a wide range of topics within the subject of contractor management. Thus, this marked the genesis of the book *Resilient Contractor Safety Management*. This is our first pass at a very involved and comprehensive subject. I fully expect there to be future versions or updated editions of this material. I think the subject is broad, important, and has never before been discussed in the way we cover it here.

Enjoy it! You're holding in your hands the fruit of labor of some of the smartest, most experienced, and passionate contributors in the contractor management space. Their collective wisdom in this book, stemming from hundreds of years of combined experience, is worth its weight in gold. And I am proud to assemble these esteemed scholars and practitioners to bring you this book.

Dr. Simon Goncharenko, PhD, CSP, CSHO
Midway, TX – USA

Foreword

Imagine a world where an operating model involving contractors was not available for organizations. A world where there was no reliance on skills and expertise outside of an organizations' direct pool of employees, no external specialized services available to engage at the drop of a hat, a world where resourcing and supply chain challenges were on the organization to bear, where flexibility and adaptability gains were not available as projects extend deadlines or where scope creep muscles its way in, or where cost efficiencies synonymous with contractor engagement models were not realized.

Thankfully, this is not the reality for organizations around the world, as reliance on contractors to deliver services, products, or projects is a key element of most organizations' operating models, with several industries becoming increasingly reliant on contractors to help them deliver for their customers.

But among all the efficiencies enjoyed by organizations there also lie some challenges. Alongside each of the benefits of contractor engagement lives a set of operating risks that need to be closely managed. And sitting right at the top of those risks is workplace safety.

For decades, organizations have grappled with questions like:

- How do we ensure our contractors are operating safely?
- How involved should we be in the safety management practices of contractors?
- What is our level of responsibility when it comes to the safety of contractor employees? or
- Could we be held liable if a contractor employee is seriously injured?

These are all big questions and because the operating contexts and risk profiles between organizations differ from company to company, industry to industry, and country to country, there are no black and white answers. Like so many things in the workplace safety realm, the answers live in the grey.

With that said, though, there are some fundamental underlying approaches to contractor safety management that are shared among those who are leading in the space and that must be prioritized. These include having solid procurement processes, forming trusting client–contractor relationships, establishing clarity in accountabilities, maintaining transparency, and building resilience.

The client–contractor relationship is a partnership, and like any strong partnership, trust, accountability, transparency, and resilience are all key ingredients to success. However, the difficult task for the safety professional, CEO, project manager, or contract manager lies in figuring out how to apply these elements to their contractor management models and practices. But thankfully, this is where Simon steps in with this wonderfully practical and insightful book, providing a roadmap that can be followed by any organization, anywhere in the world, pulling from his own vast experience, academia, and the experiences and insights of others who are leaders in the space of resilient contractor safety management.

Simon and I share some common core values, and I respect Simon enormously as a safety leader. So, when Simon asked me to write this foreword, I was humbled and excited at the opportunity, but I was more excited to be part of something that has been long needed in the safety profession.

There is so much conjecture and confusion around contractor safety management here in Australia where I work, and differences in approaches seem to be largely based on the personal experiences of safety managers and not a lot more. Sure, the safety regulators provide some useful resources, as do emerging tools in the various safety approaches that have been gaining traction in recent years, but there lies a very apparent gap in the absence of a resource that provides a tried and tested roadmap to effective contractor safety management – or contractor safety partnering – and that is where Simon's book steps in to address this gap.

Resilient Contractor Safety Management is a must-have for the toolkit of every safety professional and business leader and has the potential to aid a global audience given the shared contractor safety management challenges faced by organizations around the world.

Simon's book pulls together decades of experience in contractor safety management from a range of experts in the field, together with academic research and authored contributions from leaders who have already put in the work, made the mistakes, extracted the learnings, and then generously offered those learnings to you, the reader.

So, congratulations for purchasing this book. You won't regret it, and either will your contractors. *Resilient Contractor Safety Management* is the way of the future.

Tim D'Ath
Safety Executive and Author of Humanising Safety – A Four Step Approach.

Acknowledgments

No one exists in a vacuum, and the story of the birth of this book is no exception. It was shaped by more individuals than can be mentioned here, but that does not mean that at least some attempt to name a few should be foregone altogether. So, here it goes.

First and foremost, I am grateful to my family for their patience, graciousness through long nights and weekends when this book was assembled, and their encouragement and inspiration along the way. I always said that I married way above my grade, and Alice, my lovely bride of 27 years, continues to amaze me with her erudition, understanding of people, and innate ability to break the most complex issues into the simplest terms. I am so grateful for her wisdom through the duration of this project.

Second, I am having the greatest time leading one of the best teams I have ever had at Veriforce, the Expert Services. These individuals are all unique in their specialized training, experience, and expertise, but united by their incredible sense of humor and humble ability to not take themselves too seriously. It's been a fun ride with them, and our conversations, interactions, and problem-solving endeavors together made a significant contribution to this volume.

Third, the larger Veriforce team, some of whom decided to join this project, has been a blessing as well. Their unique perspectives made this resource unlike anything else on the market today because their expertise and capabilities are second to none. Don't believe me, just read their chapters, and you will see.

Then, there is the largest global client network, consisting of approximately 3,000 current Veriforce clients in 140 different countries, offering an unprecedented opportunity for learning. My face-to-face and virtual interactions with many of them certainly played an important role in the knowledge and experience coming through these pages. Some of them even joined the ranks of contributors to this book, so you can hear straight from their own perspective.

Of course, I cannot ignore another powerful network that both contributed to and is positioned to benefit from this volume, the 130,000 contractors managed by Veriforce. To every contractor who ever spoke up, stopped a job, and changed a life, I say thank you and acknowledge your contribution to this volume.

Last but not least are the workers who show up, adapt, learn, grow, make errors, learn from them, become better, and are resilient. You are the reason for this book. Our ultimate aim is to help you go home better than you came to work. Better equipped. Better supported. Better prepared to work safely.

Acknowledgements

1 The Case for Indispensability of Resilient Contractor Safety Management Today

Simon Goncharenko

1.1 INTRODUCTION

Though this project was among the most important investments that British Petroleum (BP), as a corporation, was making for the future, some felt like the Macondo, one of the deepest wells drilled at the time, had been a cursed endeavor from the start. The well was a three-mile-long tube of cement and steel that had been burrowed into subsea rock in the weeks before. It was one of the industry's most important new efforts to find oil in the deep waters off the southern coast of the United States.

As is characteristic of the industry, of the 126 personnel on the rig, only six were BP employees. The remainder were contractors from Halliburton, MI-SWACO, Transocean, and others. So, from the personnel perspective, it was a typical project, the progress of which largely depended on the ability of the contractors to do their part. This is the reality of how business is done today.

Together, these contractors had battled for weeks to control wild kicks of gas and to adapt to a series of setbacks doled out by the complicated and unpredictable well. Stress and conflicting mandates to drill as quickly, safely, and cheaply as possible resulted in many wrong decisions. These decisions, combined with failed sensors, automatic shutoffs, and other equipment malfunctions, resulted in the Deepwater Horizon explosion at 9:56 pm on April 20, 2010.

Known today as one of the most significant modern industrial incidents, the long-lasting and far-reaching consequences of the Deepwater Horizon catastrophe were truly devastating. The full costs of this dreadful explosion are hard to measure, though it immediately claimed 11 lives, injured 17 more, spewed 134 million gallons of oil into the Gulf of America over the course of 87 days nearly a mile below the surface, and amassed a hefty $70+ billion price tag, including spill response and clean-up, fines and penalties, and litigation. In the ensuing days and years following this disaster, there have been numerous investigations, finger-pointing, and reports

DOI: 10.1201/9781003679448-1

that identified issues in the areas of corporate responsibility, business ethics, leadership, decision-making, communication, and contractor management.

And while the point of this chapter is not to get to the bottom of all the contributory causes of the Deepwater Horizon incident, the significance of the interplay between BP as a hiring client and its respective contractors cannot be overestimated. Aside from all other significant triggers of this event, the backstory of BP culture and how that culture impacted its interactions with all the contractors around the world cannot be ignored because long before that fateful day, it set the tone for what took place on April 20, 2010. After all, even at the time of the event, BP contractors made up 95% of the workforce on the rig. So, it goes without saying that contractor management played perhaps one of the key parts in what many consider today the most significant industrial disaster of the century.

Resilient Contractor Safety Management is paramount to the success of many, if not most, organizations today. Contractors can make you or break you. Thus, the reason for this book.

An internet search conducted in early 2025, prior to the writing of *Resilient Contractor Safety Management* revealed that there was no other resource like it available at that time. Sure, there are a few books that have similar titles, but none of them has the breadth, the cross-industry reach, and depth of expertise of this manuscript. It is a combination of academically researched theories and decades of practical field-tested experience, authored by a wide range of individuals from all walks of life.

The aim of this book is to help different types of leaders across all industries, from executive level to mid-management, to front-liners. *Resilient Contractor Safety Management* can offer guidance for operational leadership, procurement personnel, compliance function, HR experts, and health and safety professionals alike. Given the highly practical nature of this book, its guidance can be helpful whether you read it cover-to-cover or select specific chapters for reference.

1.2 CONTRACTOR SELECTION

Each company that relies on contracting work for its success must prequalify and vet those contractors on the basis of some clear and defined parameters, like cost, quality, safety, scheduling availability, insurance, etc. Sometimes the vendor selection criteria are clear-cut, and it is easy to award or reject contractors based on their risk levels. At other times, however, due to a wide range of reasons – such as remoteness of the work, the highly specialized nature of the contract, special insurance requirements, or many other criteria – a riskier contractor has to be selected, and the hiring client better be prepared for conducting deeper scrutiny. If this is an area where you need further direction or could use some extra data, don't miss Karl DeLooff's chapter on Contractor Selection (Chapter 2).

Of course, the prequalification and the risk-based vetting are only the beginning of the contractor selection journey. Once that step has been completed, the next element is defining and agreeing on clear scopes of work, roles, and limits of authority. Crossing this bridge early can prevent unnecessary conflict and unpleasant future conversations on scope creep and other related matters. Samuel Rayborn's chapter, Better Contractors Selected When Safety and Procurement Collaborate (Chapter 3), can be a valuable resource for furthering the conversation on proper terms and conditions of contractual definitions and obligations.

At this stage of the journey, it is also important to verify that contractors have qualified personnel for the specific tasks, jobs, and hazards involved. If generic experience will not do in your case, vet your contractors against the specialized background, training, qualifications, and certifications. The more work you do at the front end of the contract, or before it, the easier your life becomes after the signing of the contract.

1.3 THE NAME OF THE GAME IS PARTNERSHIP

Why do you hire contractors? Because they are uniquely qualified to do what you cannot do at the same cost, schedule, or with the same accuracy, precision, quality, etc. Yet, in hiring contractors, you incur a level of risk by attempting to merge your culture with theirs while still maintaining the legal separation that will steer you clear of co-employment clauses and such. So, how do you do all that and do it successfully? The simple answer to this complicated question is partnership. If you partner with your contractors, treating them with respect, honesty, and openness, they will do their utmost to satisfy your requirements. Ron Gantt's chapter, Contractor Partnership (Chapter 4), is filled with very insightful and practically applicable recommendations on this subject, and you would be remiss not to read it. The key contributing factors in his chapter are absolutely indispensable to a lasting and prosperous relationship with your vendors. Start with his chapter and build on it the other components that may be unique to your situation, context, or geography.

In the same spirit, Chris Detillier's chapter, Client–Contractor Collaboration and Ways to Achieve It (Chapter 5), adds to this area by providing some additional considerations and questions that need to be addressed in the area of client–contractor partnership. And then Josh Ortega's chapter titled Continuous Improvement Requires Employee and Contractor Involvement (Chapter 6) puts a cherry on top of this subject in an erudite and practical manner. The material offered by these three individuals is practical, field-based, and easily implementable. A combination of the chapters by Gantt, Detillier, and Ortega would give you a solid roadmap to positive client–contractor interaction.

1.4 CULTURE, CULTURE, CULTURE

Another extremely consequential element of contractor selection is gauging the level of alignment with safety culture. This term and the concept it represents may be entirely new to you. I will provide a quick overview of what it means here while also recommending my other chapter (Simon Goncharenko) entitled Charting a Just Course: Where Fairness Meets Accountability (Chapter 7) for a deeper dive into the subject. What is that just course? This is a culture where mistakes and human errors are not punished indiscriminately but are learned from. After all, we are all human and make errors all the time. Yet, at the same time, the intersection of fairness and accountability brings understanding that willful violations of known safety policies that endanger lives should not be tolerated without appropriate accountability.

Even a quick internet search reveals a large number of competing definitions of safety culture on one side and an even larger number of voices on the opposite side arguing for utter lunacy of the concept. After all, this latter group says, we don't have a finance culture, or an HR culture, so how can we possibly attempt to capture the absence of unwanted events as this made-up social construct, called safety culture? What this group argues makes sense, to a point. However, our constructs are as dynamic as our languages.

So, for example, the word "woke" was not even considered a grammatically correct term a decade ago and was therefore not in use at that time. But today, this once nonexistent term means a very specific set of values, worldviews, and leanings, and people not only know it and use it, but they also feel very strongly about it – one way or the other.

In the same way, the term "safety culture," while perhaps somewhat nonsensical in and of itself, may be the easiest way to identify the sum of what an organization is and does in the pursuit of safety. Safety culture is a broad umbrella that incorporates a wide range of items, such as perceptions, policies, procedures, standard operating procedures (SOPs), leadership engagement, values, communications, training, psychosocial considerations, incident management, drug and alcohol expectations, and a whole host of other elements.

The reason safety culture is even brought up in this introductory chapter is because it is very important to get it right. If we don't, people die. And no one dies if we get the "accounting" culture wrong.

But there is also a commercial reason why safety culture must absolutely be considered in the process of selecting the right contractor. As was mentioned in the introductory part of this chapter, your associations with contractors bring a degree of shared legal and reputational liabilities that come with the territory. So, proper vetting on the front end can result in saving much grief, litigious pain, and reputational catastrophe on the back end.

Okay, we established the importance of alignment on culture, specifically related to safety here. What we have not really mentioned or discussed is how you ascertain that there is an alignment. Clearly, there is more to it than simply asking about the trade partner's culture and taking their word for it.

So much more. Paul Espenan's chapter, One Question: A Window into Contractor's Maturity (Chapter 8), can be a great place to start for putting together a solid strategy of measuring or gauging a contractor's culture.

1.5 CULTURAL ALIGNMENT IS CRITICAL

What does it actually mean to have an alignment with the culture on a practical level? And what if there is no alignment with the culture, but you must hire the contractor because there are no alternatives? These are all good questions, and they have good answers.

In fact, both of the questions above can really be answered together. Here's a practical implication – whether you do have an alignment or you don't, as a hiring client, you can set the expectations for the right culture for all your contractors. As such, those whose internal cultures are aligned with yours can enjoy a mutually beneficial relationship that encourages growth, sharpening, and improvement. Contractors who are not aligned, however, can still be expected to adhere to your standards. Here, a specific and detailed plan will need to be worked out on what your expectations are and how this contractor can meet them.

Your expectations may be for your employees and contractors to meet the minimum regulatory requirements of the appropriate federal, state, and local jurisdictions. If this is you, you are not alone. Many clients share these expectations.

Or your expectations may go above the minimum regulatory requirements, incorporating industry's best practices, ISO standards, and other local, regional, or global codes, procedures, and methods (ANSI, ASTM, NFPA, API, VPP, ILO, IOGP, etc.). Many forward-looking and proactive organizations are on this path today. They look at risk in a more holistic and comprehensive way than what is only required by current regulations.

Whatever your expectations may be, there is nothing abnormal about requiring your contractors to meet your expectations. It is your prerogative, and if they want to do work for you, they will need to choose whether this relationship is worth the additional costs of compliance.

Another practical element in the area of the alignment with safety culture is the expectation that your contractors have for you. What do they expect from you? They expect you to provide onboarding for their personnel, offer site-specific orientations, and clarify your expectations for what success looks like on your jobs.

1.6 CONTRACTOR INDUCTION AND TRAINING REQUIREMENTS

The next element of a successful partnership between clients and contractors is the proper induction of your contractors. When you conduct the induction of your trade partners, you have the opportunity to impart to them what separates you from all the other hiring clients. The following section contains just a small sample of the types of questions and issues that you may want to consider including in your contractor induction:

- What makes you, you?
- Is there something unique about your culture? Introduce them to it.
- What are the most important elements of your culture? And which parts of it do you expect your contractors to embrace?
- How are your sites different, unique, or specialized?
- What else do you contractors need to know about your sites?
- What do contractor employees need to bring or not bring to your sites?
- What are your nonnegotiables?
- List your expectations, especially if they are above the minimal regulatory requirements.

Contractor induction is followed by contractor training expectations. Contractor training is an indispensable element for a successful partnership. In the training expectations of your contractors, you will start with the induction described earlier and build on that foundation. Your organization undoubtedly has a training matrix for your own personnel. Which elements of that curriculum do you think should be applicable to your contractor employees?

Setting the minimum training expectations for your contractors is only half the battle. The other half is verifying that the content and the delivery of the training chosen by your contractors actually meet your expectations. Otherwise, how would you know if their fire extinguisher user training is fully online and never touches a fire extinguisher? Or if their Control of Hazardous Energy or Lock Out Tag Out (LOTO) training is 2 minutes and 45 seconds long and delivered in an e-learning format? The last question was an actual discovery made in the contractor training as a result of an incident investigation. You must remember here that the reason you require training from your contractors in the first place is not for the sake of them having a checkmark for training, but for the sake of them learning and understanding what to do and not to do, in order to go home safe, primarily, and in order not to damage the property or equipment, secondarily.

For more on the contractor training elements, check out Ron Gantt's chapter titled Getting Better When Nothing Happens: Proactive Learning for Clients and Contractors (Chapter 9). Ron's perspective is unique but important for you to consider.

1.7 MANAGING HAZARDS AND EMERGENCIES

The next important element of partnering with your contractors is ensuring that they share your understanding of what constitutes a hazards across your sites, facilities, and locations. This is related to the previous subject of training, as you will need to make sure that your vendors understand, share, and are aligned with your identification of hazards, mitigation of the same, and erection of robust systems and protocols that will prevent these hazards from hurting or killing people, damaging facilities, or destroying equipment. In addition to training on hazard identification, you will also need to educate, empower, and encourage robust and consistent assessment of risk on your jobs. Josh Ortega has an excellent chapter on this subject, titled Contractor Hazard Identification and Risk Assessments (Chapter 10).

Being prepared for emergencies is an important part of doing business. Whether it is violent thunderstorms, hurricanes, tornadoes, workplace violence, or terrorism, knowing what to do, how to mitigate, and seeking help can make a difference between life and death. Though we often take it for granted that everyone has an emergency action plan, it is imperative to require that your contractors have one, that it is specific enough, that it is updated regularly, and that it includes notifying you on a timely basis. For more regulatory requirements and best practices on this important element, see Rick L Ingram's chapter on Contractor Emergency Action Plans (Chapter 11).

1.8 MONITORING YOUR CONTRACTORS

What comes next? After qualifying, selecting, onboarding, and verifying all the right programs and protocols for your contractors, do you just release them to work on your locations? Not exactly. Next comes contractor monitoring, in order to validate that work at your locations is performed as planned and as safely as promised.

Trust but verify is a well-known concept, and it is certainly applicable here. There is quite a bit that can be verified in this area, from auditing the contractor's programs – be they safety, environmental, security, insurance, risk management, etc. – to assessing the implementation of these programs, examining the adherence to program and regulatory requirements at work locations, and measuring internal climate through perception surveys, etc. There are many ways that contactors can and should be monitored. In this process, the following represents just a small sample of the questions that should be considered by all hiring clients at some point:

- How do you audit or assess your trade partners?
- What do you look for in audits?
- How often do you conduct assessments?
- How do you track the closure of gap findings identified through these audits?

These and many other questions like them must all be answered, in order to establish a flourishing and productive contractor assessment strategy. Josh Ortega's chapter titled Contractor Assessment Audits: A Strategic Function in Mature Contractor Management Systems (Chapter 12) is a great place to start with all the tips and hacks from a seasoned professional (who was once referred to as a walking encyclopedia by his supervisor).

1.9 CONTRACTOR FIT-FOR-DUTY PROGRAMS

Continuing along the lines of the trust but verify principle, all hiring clients rightfully expect their contractor employees to show up to their job locations unimpaired by the effects of drugs and alcohol. Part of that expectation is aligning with your contractors on their drug and alcohol programs. When and how do you expect them to verify their employees' fit-for-duty status?

- Pre-employment? Most include this.
- Post-incident? Most include this also.
- Random testing? Most programs require this element as well.
- Reasonable suspicion? This also happens to be a common element in most programs.

For more discussion of this subject, see my chapter (Simon Goncharenko) titled Drug and Alcohol Tests for Employees and Contractors (Chapter 13).

There is another aspect to the fit-for-duty programs and expectations that needs to be addressed here. This element, referred to as case management, addresses how contractors manage the injury response and treatment of their personnel. While in theory, this may seem like more of a personal issue for the contractors and one that should perhaps be outside of the scope of the hiring clients, in practice, however, a different picture emerges. There is much more to case management than how quickly the employee returns to work after the incident. This is certainly at the center of case management, but a robust program begins way before the injury takes place, manages the organizational response to injury, promotes employee morale, and is beneficial for the workers and their employers at the same time. At the center of it is the idea that work is restorative and promotes quicker healing physically, emotionally, and psychologically. For more on the importance of properly managing return to work after an injury and its multiple benefits, see my (Simon Goncharenko) chapter titled Return to Work Reimagined: A Holistic Framework for Employee Well-being and Business Success (Chapter 14).

1.10 INCIDENT AND NEAR-MISS INVESTIGATIONS

Incidents and injuries happen in our workplaces because they are staffed with fallible people. What does it mean? It means that our employees, their managers, and the executives make mistakes. We all do. We are all fallible people. To err is human.

What's worse is that we can't exactly predict the next time that we are going to make an error. If we could, we wouldn't make that error. So, if we can't predict every error with 100% certainty, we can't prevent all incidents with a 100% success rate. What's the next best thing then? Preventing what we can and building controls in place that can help us recover by failing safely for what we cannot prevent. So, this way, when incidents do happen, they do not kill or seriously injure our people. Then, we do our due diligence by carefully investigating every single incident in order to identify any potential systemic gaps that need to be addressed.

Do your contractors conduct incident investigations with the purpose of learning and sharing the learnings across your locations and their larger team? Are their investigations up to par and do they satisfy your expectations? How do you know?

For more on how to ensure robust incident investigations, you need to see Karl DeLooff's chapter titled Incident and Near-Miss Investigations You Should Expect from Your Contractors (Chapter 15). Karl provides you with a textbook example of the do's and don'ts of proper investigation. While you are thinking of incidents, have you noticed that there is a renewed interest in serious injuries and fatalities (SIFs) in this space today? The reason for this is the realization that what causes sprains and strains is usually not the same thing that kills and maims our workers. So, while many larger organizations have succeeded in significantly lowering their recordable incident rates, often bringing them below 1.0, these same organizations continue to struggle with eliminating fatalities and other serious events. For more on SIFs, where the conversation is today, and where it is likely going in the near future, see my chapter (Simon Goncharenko) titled When Low Injury Rates Lie: The Hidden Face of SIFs (Chapter 16).

Then, after you read DeLooff's and my chapters, check out Chris Gafford's Corrective Actions as the Focus of Incident Investigations (Chapter 17). Chris is a practitioner who has been around long enough to not only see what should and should not be done, but also pick some helpful tips and tricks that you will undoubtedly benefit from. You will enjoy his West Texas style coming through his writing too, as he is a native of Midland, one of the hottest oil patches in the U.S.

We live in the age of technology, and it can and should certainly be leveraged in your contractor management practice. There are plenty of choices on the market today that you can select to assist you with either navigating individual aspects of contractor management, like incident investigation via KPA, or handling the entire suite of selection, qualification, onboarding, training, and auditing your contractors, by the likes of Veriforce, the largest global provider of such solutions, whose slogan is "We Power End-to-End Contractor Risk Management." For more information on incorporating technology in your contractor management, see my chapter (Simon Goncharenko), titled Leveraging AI and Other Technology for Smarter Contractor Management (Chapter 18).

1.11 NICHE AND HISTORIC CONTENT

Last but not least, this book also brings some uniquely specialized considerations of managing operator-qualified (OQ) contractors who work on pipelines, which, in the United States, is regulated by the department of transportation (DOT) and administered by the Pipeline and Hazardous Materials Safety Administration (PHMSA). Thomas N. Meek, Jr., who spent close to three decades in the pipeline industry, working his way up from technician to the Vice President of Compliance Operations, is by far one of the most knowledgeable and qualified individuals to author a chapter on this topic. So, I am honored to have him join our project, and you can benefit from his expertise by carefully considering his chapter titled Qualifications of Contractor Pipeline Personnel (Chapter 19).

Another chapter in this book that would fall within the uniquely specialized consideration would be my chapter (Simon Goncharenko) titled The Current of Change: Challenges and Growth for Utility Providers and Their Contractors (Chapter 20). While the chapter was written in the United States, it considers other geographies and spells out the challenges faced by hiring clients and contractors within this important group of industries currently experiencing tremendous strains from the demands placed by electric vehicles, data centers, and AI.

What makes the book in your hands truly unique and one of a kind is the breadth of its content. Those who have lived at least a few decades usually recognize their connection and their debt of gratitude to those who came before them. Let's face it, we are who we are today because of the accomplishments, discoveries, struggles, and victories of the previous generations.

So, connecting you, the reader, to the generations that came before us is Jim Wetherbee's chapter. Captain Wetherbee, a US Naval Aviator who became an astronaut and later served as Director of Flight Crew Operations at NASA, authored a chapter that provided a historic overview of the contractor assurance process in the construction of the International Space Station. If you read anything about it, you can understand what a feat that project ended up being in its own right, involving the coordination and management of thousands of contractor companies from all over the world. So, an insider perspective is a unique treat for the reader and should be savored by examining Wetherbee's chapter titled The Role of Effective Assurance Process in the Construction of the International Space Station (Chapter 21).

Finally, 2024 brought a monumental shift to the interplay between government and industry in the United States, with potential consequences for the rest of the world, when the Supreme Court struck down the Chevron Deference doctrine. For a detailed discussion of the case, its implications so far, and what is expected, read the last chapter of this book, authored by yours truly and titled The Impact of Chevron Deference on American Clients and Contractors (Chapter 22).

1.12 BEST USES FOR THIS MATERIAL

As you can see, this is quite an amazingly comprehensive manuscript compiled by the brightest and best minds across various industries. At the time I am writing this, there has not been a product on the market that attempted to deliver what is in this book. So, be sure to check it out.

Read a chapter that most urgently addresses the needs or pain points of your organization today. Or you can read the entire volume; we purposefully made our delivery easy to consume. After completing this book, why don't you buy a copy for a friend or a colleague?

And may I suggest that you urge your contractors to obtain a copy or a few copies of this book and familiarize themselves with its contents. It could be the single most important thing you do to help them progress, mature, and improve their internal systems, processes, and protocols. I guarantee they will thank you for this later.

If you are an academic or operate in the academic circles, consider introducing this book into the required reading for your undergraduate- and graduate-level business management, safety management, and procurement/supply chain management degrees. The academic integrity of this resource is backed by the fact that two of the authors have earned Ph.D. degrees. However, at the same time, this manuscript also delivers where most purely academic resources often fail, which is in the area of providing tried-and-true, field-tested practical application recommendations. So, go ahead and add it to the required reading list. Your students will thank you for it later.

And now, enjoy!

Dr. Simon Goncharenko, Ph.D., CSP, CSHO
Midway, TX
July 2025

BIBLIOGRAPHY AND FURTHER READING

Lustgarten, Abrahm. (2012). *Run to Failure: BP and the Making of the Deepwater Horizon Disaster*. New York, NY: W.W. Norton & Company, Inc.

2 Contractor Selection

Karl DeLooff

2.1 INTRODUCTION

Selecting the right contractor is essential to achieving project success. Past performance remains the best predictor of future performance. Contractors operate within a framework of formal systems, tribal knowledge, and personal experience. Those with a history of consistent and predictable performance are more likely to meet project objectives, including safety, reliability, and profitability.

Contractor selection is an exercise in risk reduction. No matter what industry you operate in, sourcing contractors carries inherent physical safety risk. This discussion excludes virtual risks associated with software development and similar projects. Contractors that provide skilled labor at risk must be mature and have well-developed systems of operation with a continuous improvement expectation. Another assumption is that the evaluator is the project owner or general contractor, who is both assuming risk and assigning risk during the contracting process.

Projects inherently contain known, unknown, and other latent risks. The various risks can be grouped as follows:

- Financial risks include the inability to fund the operations pending the payment of invoices.
- Subcontracting risks that replicate all the risks to a contractor.
- Design risks, including unsuitability for purpose.
- Supply chain risks, including disruptions and damage to single-point-of-failure equipment.
- Security risks, including theft and vandalism.
- Quality risks, including rework and warranty cost.
- Schedule risks, including delays and conflicts.
- Safety risks, including injury prevention and case management.

Financial risks can be discovered and controlled by a fiscal evaluation and by listing, as a line item, the cost of a performance bond. A monthly release of lien system should be an element of the contractual language for the contractor's suppliers and subcontractors. Failure to pay or to pay on time is a common issue that has significant reputational risks to the owner, and there must be a system of controls and routine monitoring at least monthly. Dunn and Bradstreet, as well as a host of other vendors, can provide assistance in this area.

Contractors must own the entirety of their subcontractors, including selection, leadership, and management. Contractors who are either unwilling or unable to provide effective ownership of their subcontractors must be rejected or prohibited

DOI: 10.1201/9781003679448-2

from directing subcontractors. Effective ownership does not permit excuses or "passing-the-buck."

One of the easiest ways of determining if avoiding responsibility is a pattern for any given contractor is by interviewing previous project owners, subcontractors, and suppliers. The subcontractor selection process that the contractor employs can be valuable information. The contract negotiation stage, with strong language over subcontractor ownership, will also reveal the normal operating pattern.

It has often been said that the best project engineer is the one you have not used yet. Every design contains trade-offs that have both positive and negative consequences. The impact of those decisions will contain latent conditions that will only be manifested under unpredictable conditions present in the future. Other impacts will manifest themselves during construction, during commissioning, in the first months of operations, during the first or second major maintenance cycle, and in other long-term stages and events. Making design changes during the initial design process escalates the cost more than linearly as the project design progresses and geometrically during the completion of the physical work. Tracking design change orders from previous owners is the best way to determine how well the designers/engineers translate the functional outcomes for the application.

The supply chain is another risk that must be qualified and potentially quantified. Certain items will have a long lead time or represent single points of failure. These items will be on the critical path for the project schedule. Although other components may be considered noncritical path items, they too could have an out-sized impact on the success of the project. Those items should be understood from both a logistical and a cash flow perspective. Early procurement and storage may solve one issue, but security, maintenance, degradation, and shortened warranty durations may create their own issues. Contractors will differ on how to address the supply chain risk. Contractual language must be clear about who has responsibility for supply chain risk.

The security of the site, materials, equipment, and the solvency of the contractors is a risk that must be addressed in the contract language and actively administered on the site. Some security issues are based on the interaction between the site and the local community. Other security issues are insider-based, and the staff on the site are part of the overall risk profile. This may include theft of materials and equipment or invoice fraud. Contractors should be able to provide builder's risk or general liability insurance, as appropriate to the project. There should be a series of controls covering physical security, monitoring for unauthorized access and vandalism, inventory management of materials and equipment, and oversight of project finances.

Quality and craftsmanship are risks that contractors can flex to meet project financial and schedule goals. The goal is always "Right the First Time" or first-time quality – the right part/material should be incorporated into the finished work to meet project specifications for the long-term functioning of the project. The obstacles to the first-time quality could be many, including undertrained staff, inferior quality parts/materials, poor procedures, skipping intermediate hold points, inadequate tooling or measurements, poor or absent expert supervision, and a host of other factors. Rework is costly for the contractor, frequently unsafe, and the final product is often substandard.

An additional complication with rework is that often, contractors will be seeking change orders. Project owners may demand an extended warranty. Warranty costs may be high and subject to a dispute between the contractor and owner. Interviews with previous project owners will likely be the best source of information to determine quality risk.

Project schedule drives cost, price, and compliance. Electrical power generation construction, for example, requires meeting both regulatory and grid schedules. Schedule slippage may result in a power plant that cannot operate for an extended period of time – sometimes years. Managing schedules, including weather, subcontractor, and supply chain risks, requires significant financial and intellectual resources.

There will be significant differences between contractors in the level of sophistication and effectiveness in managing project schedules. Compressed and recovery schedules often have adverse safety and quality impacts – there is insufficient time for training, competence verification, and quality inspections for the work incorporated into the next step. An example is rebar before a concrete pour. Interviews with previous project owners will likely be the best source of information to determine schedule risk.

Contractor selection controlling for safety involves assessing known and unknown variables to predict future performance. While some factors are quantifiable, others require subjective judgment. Easy measurable metrics, such as safety incident rates, may not always strongly correlate with overall contractor effectiveness. The best contractors adopt structured methodologies for continuous improvement. However, penalizing contractors for experimental efforts – such as testing new processes or equipment that inadvertently result in a reportable safety incident – can discourage innovation. Although safety failures must be addressed, fostering a culture that values learning and improvement is essential.

There are vendors that provide safety evaluations of contractors. These systems can provide input, but they are not the fire-and-forget weapons to base contractor selection upon. These systems are based on easily identifiable metrics, such as Occupational Safety and Health Administration (OSHA) injury rates, part of the US Department of Labor, which may or may not include the context of the injury or violation. For example, a company that takes ownership of the incident and the results will improve its systems, procedures, and may identify latent factors that remain hidden until that point. Another company may blame the injured and be set up to repeat the event. Both would have a similar score from the evaluator of the metrics, but one company would be superior to the other in the future. Learning how to discriminate between the two is the subject of this chapter.

There are several versions of safety and quality management systems, including the International Standards Organization (ISO) 45001 standard for safety systems and ISO 9001 for quality management systems. The United States OSHA has the Voluntary Protection Program (VPP) and Safety and Health Achievement and Recognition Program (SHARP), which deserve respect for delivering on a low-consequence, worker-centric safety culture.

Much of the criteria presented below are measurements of company culture. There is a discussion about whether safety culture is distinct from other company cultures.

There is a definite company culture that is partially built and partially organic. The better cultures have the characteristics of good leadership and highly effective teams. A project owner or general contractor will have a difficult task attempting to directly and objectively measure leadership quality and team dynamics among hundreds of teams whose members may vary depending upon the scope of work. Consequently, the following measurements are more obvious and quantifiable, even if in a subjective question for interpretation.

2.2 CONTRACTOR EVALUATION CRITERIA

A structured evaluation system ensures objective contractor selection. The following criteria provide a robust framework for assessment:

2.3 PAST PERFORMANCE

- Experience modifier rate (EMR)
- Total recordable incident rate (TRIR)
- Lost time injury rate (LTIR)
- Regulatory or legal violations (OSHA, EPA, permits, etc.)
- Lawsuits and legal disputes
- Schedule adherence (liquidated damages history)
- Budget adherence (change orders and cost overruns)
- Warranty claims and rework history
- Employee training records
- Skills verification records

A contractor's historical performance provides insight into their ability to manage safety, budgets, schedules, and compliance. While no single metric should determine selection, identifying trends across multiple categories can paint a more accurate picture of contractor reliability.

2.4 SYSTEMS AND PROGRAMS

- Employee and contractor safety manuals (aligned and integrated with project requirements)
- Onboarding and offboarding programs for employees and contractors
- Training programs for skills development
- Competency verification programs
- Formal mentoring programs with progress tracking
- Lessons Learned documentation and integration process
- Continuous improvement and innovation initiatives
- Corrective and Preventive Action (CAPA) systems
- Leadership development programs for all staff levels

Structured programs demonstrate a contractor's commitment to maintaining consistency and improving operations. However, the effectiveness of these programs in practice is a more critical measure than their mere existence on paper. There must be evidence of positive impacts on outcomes over time.

2.5 KEY INDICATORS FOR EVALUATION

- Documented competency verification failures (demonstrating rigorous standards)
- Safety professional competency evaluation (beyond certification)
- Leadership development tracking and training participation
- Mentorship program effectiveness (leading to competency validation)
- Employee turnover rates by role and organizational level
- Documentation of corrective, preventive, and innovative actions from identification through implementation
- Safety inspection records and hazard recognition logs
- 360-degree referrals from past clients, subcontractors, and suppliers

2.6 EVALUATING PAST PERFORMANCE

The most accessible past performance data typically includes OSHA regulatory safety records, such as TRIR and LTIR, and insurance-based EMR ratings. Comparing these figures to Bureau of Labor Statistics (BLS) benchmarks for relevant NAICS codes provides context. However, these numbers disproportionately favor larger firms, which benefit from more administrative staff and diversified risk pools. Conversely, smaller contractors may appear riskier due to the statistical sensitivity of their incident rates.

Regulatory violations and lawsuits further inform contractor evaluation. OSHA's establishment search tool can reveal prior safety infractions. Repeated violations, particularly those related to systematic safety issues, may indicate leadership deficiencies. Similarly, frequent legal disputes may suggest ongoing compliance or operational struggles. Assessing these factors in context ensures a fair evaluation of whether these concerns reflect isolated incidents or broader organizational problems.

2.7 ASSESSING OPERATIONAL EFFECTIVENESS

Operational performance directly correlates with contractor reliability and safety outcomes. Key indicators include:

- Adherence to project schedules and budgets
- Frequency and nature of change orders
- Warranty claims and rework rates
- Design and execution quality (for EPC firms)
- Training records showing skill progression and leadership development
- Employee retention and career advancement within the organization

Contractors that demonstrate workforce stability and invest in skill development are more likely to uphold strong safety standards. Examining historical employee progression from craft roles to leadership positions signals a commitment to long-term development.

2.8 EVALUATING SYSTEMS AND PROGRAMS

A contractor's written policies and procedures set the foundation for performance, but their implementation determines effectiveness. Key considerations include:

- Frequency of policy updates to incorporate lessons learned
- Integration of field knowledge into formal documentation
- Alignment between documented procedures and actual worksite practices
- Effectiveness of employee and contractor onboarding programs
- Mechanisms for monitoring and improving safety culture

Onboarding programs should ensure all personnel understand safety expectations, the scope of work, and procedural compliance. Additionally, a contractor should have a clear process for managing subcontractors, ensuring all workers adhere to established safety and operational standards. The employee onboarding process should be evaluated for completeness and depth.

2.9 TRAINING AND COMPETENCY VERIFICATION

A contractor's training program should extend beyond regulatory compliance to verify practical skill proficiency. Key evaluation criteria include:

- Hands-on competency assessments rather than relying solely on certificates
- Formal mentoring with structured goals and evaluations
- Integration of lessons learned into ongoing training updates
- Leadership training tailored to different organizational levels

A robust competency verification system should include records of failures, demonstrating rigorous standards. If 100% of workers pass competency assessments without retests, the system may function more as a formality than a genuine evaluation tool.

The verification of skills and competencies involving high safety risk and critical quality standards is the most important. For example, high-voltage cable splices/terminations and energized electrical work are high-consequence skills from a quality and safety perspective.

2.10 LEADERSHIP AND WORKFORCE STABILITY

Strong leadership is fundamental to contractor success in safety, quality, and efficiency. Important indicators include:

- Leadership training participation and career progression records
- Stability of key personnel, particularly in safety-sensitive roles

- Turnover rates exceeding industry norms (>15% for stable office roles, >20% for traveling personnel)
- Evidence of cohesive teams and a well-integrated safety culture

Frequent leadership changes or reactive policy shifts may indicate internal instability. In contrast, an incremental, continuous improvement approach suggests strong leadership and sustainable organizational growth.

2.11 SAFETY INSPECTIONS AND HAZARD RECOGNITION

Safety inspections should evaluate not only regulatory compliance but also risk management effectiveness. An effective safety program:

- Focuses on risk mitigation beyond personal protective equipment (PPE) compliance
- Ensures job safety analysis (JSA) processes are thorough and worker-driven
- Tracks identified hazards from recognition through resolution
- Integrates findings into formal corrective and preventive action programs

Successful contractors use safety inspections as continuous improvement tools rather than simple compliance exercises. The main difference between an effective and an ineffective safety inspection program is informed intentionality. The inspection criteria are based on the task, the control measures, previously communicated standards, the resources to comply with the standards, and the company's leadership support of the high standards with resources – especially time and personal attention.

The number of separate safety inspections should approximate the number of separate tasks or risks identified during a pre-project risk assessment. Some tasks and risks may be similar and allow for some consolidation, but to keep the number of criteria to a manageable number and focus the attention of the safety inspector, the tasks should be discreet. One common element that should be on all safety inspection criteria is the JSA process, which includes a document, an effective safety briefing, and certification. The goal of the JSA process is a deep understanding of the tasks, risks, and effective mitigation measures. Crews that properly use the JSA process are typically the highest performing in terms of both safety and quality of work.

Evaluate 10–20 inspection documents for depth, completeness, and follow-up with the crew members. The inspection process should include a discussion with some crew members and the foreperson. You should see far more compliant and positive comments than non-compliant and negative ones. The inspection criteria should not be limited to safe/unsafe but should include at least a three-position scale: acceptable/marginal/unacceptable. The real value to the leadership of the company is in the "marginal" category, where the proactive actions can be executed. The "unacceptable" categories are reactive and should be corrected. The "marginal" area is where the true improvements that uncover latent conditions are made. These will improve the processes, worker morale, and overall safety.

Hazard recognition programs are a hallmark of a mature safety program and an effective safety culture. For a hazard recognition program to function, there must be trust and follow-through by the company from the perception of their workers. Most companies will have a program that has either limited use or mandatory compliance. The limited use cases typically start with a program rollout, accompanied by high hopes from the workers at risk, and then result in crushed hopes when there is no improvement or communication. The mandatory use programs typically have meaningless submissions, or the program is used for passive-aggressive backbiting.

Evaluating a hazard recognition program is more anecdotal than data analysis. Request submissions from 4–6 months ago for a definite time period, such as 2 weeks or a month. Read through them for a reflection of well-considered hazards. Select a handful of hazard submissions that should be improved, and track the corrective or improvement action through completion. Track what communication occurred with the reporter (or crew) and the entire company. If there is a good track record of follow-through and completed improvements, then the program is valid, and the safety culture is likely positive.

Another key marker is the continual improvement of the main program and policy documents. If the corrective and improvement actions generate a new policy document or a singular program, there will be chaos in the system, and the workers will wait until the topic loses the interest of the leadership and the process regresses to the mean. There may be a minor improvement in one area for a short duration, and then the leadership capital is expended without long-term improvement. This cycling of initiatives and programs is highly problematic and shows a reactive and poor safety culture.

An indicator of reactive leadership and poor safety culture is a continual accretion of singular documents of rules. The total volume continues to increase, and many are older than the technology that is currently in use. The overall accumulation and ignorance of the various documents make compliance impossible, and the overall safety program loses its legitimacy in the minds of the workers. Since they cannot know what applies, keep the policies and procedures applicable to the works and tasks straight, the workers will default to what seems to work and tribal knowledge – negating the policies and procedures leading to another incident or injury, investigation, and another policy or procedure. The additional document creation cycle will soon repeat the pattern of being ignored by the workers and leadership. It is a vicious circle that becomes a doom loop.

2.12 REFERRALS AND INDUSTRY REPUTATION

Past project references provide a practical measure of contractor performance. A structured referral process should include:

- Contacting owners, subcontractors, and suppliers from the last five similar projects
- Confirming whether reported safety records align with field experiences
- Evaluating warranty claims and post-warranty performance
- Assessing whether clients would choose to work with the contractor again

Frequent geographic movement and lack of repeat clients may indicate a contractor's attempts to avoid reputational consequences. Conversely, long-term relationships suggest a commitment to quality, reliability, and collaboration.

Most organizations will not document or submit to another organization their opinion and metrics of a prior contractor. Most sophisticated organizations have an internal evaluation process for assessing the prior project and the performance of their various contractors. However, these metrics are often lost during the post-project transfer process and the push to monetize the new asset. A short interview with diagnostic questions that is entirely verbal will be the only way to get the necessary diagnostic information.

The key personnel to interview:

- The new asset's maintenance manager
- The engineer of record
- The owner's project manager
- The project manager of a subcontractor

Each point of view has its own information essential for a solid evaluation. Here are some questions that need to be answered by at least one of the individuals:

- How many injuries were reported during the project? Does this seem high, average, or low for the project?
- How many near-misses were reported during the project? Does this seem high, average, or low for the project?
- How many property damage incidents were reported during the project? Does this seem high, average, or low for the project?
- How quickly were the various incidents reported? Did this seem timely to you?
- Were root cause analysis reports in-depth, and did they identify the true causes? Were the corrective actions effective?
- Did the contractors' internal processes seem smooth and mature, or did you experience chaos and difficulty?
- Did the contractor assign blame to others (owner, engineer, suppliers, subcontractors, etc.)? Was the contractor collaborative with proposed solutions?
- Were the change orders legitimate and realistic?
- What was the initial quality of the project? Were warranty items quickly resolved? Were you forced to accept lower quality, durability, or reliability than specified?
- Was the work in progress fully inspected before being covered by subsequent work? What quality documentation went missing or was unfulfilled?
- How well was the project schedule adhered to?
- How accurate and complete were the operations and maintenance manuals?
- What rework or immediate repair was required?
- Did you experience any reputational damage from the project or contractor?
- Would this contractor be on the top of your list to work with again?

- What advice would you give an organization considering using this contractor?
- What "best practices" modeled by the contractor did you adopt?

For the subcontractor:

- Were you paid on time?
- Was any injury or incident reporting suppressed?
- How well was the critical path coordinated? Were any hold-points for inspections violated?
- Would this contractor be on the top of your list to work with again?
- What advice would you give an organization considering using this contractor?
- What "best practices" modeled by the contractor did you adopt?

From the answers to these questions, make a judgment on whether to place the contractor in the 20-percentile bracket – essentially a 1–5 rating scale. The top 20% being a 5 and so on. This rating will be useful in the selection process. If you get responses that indicate poor safety practices – especially failure to report or investigate incidents – that may be an elimination factor.

Do not be too impressed with large numbers of hours worked and very low incident rates. That could mean a whole range of things, such as the contractor being good, lucky, or dishonest. A properly performed interview process will allow the data to be placed into the context of real-world performance.

2.13 CONTRACTOR SELECTION CHECKLIST WITH WEIGHTING CRITERIA

Selecting a contractor involves evaluating various performance and compliance factors. The following table provides a structured checklist with suggested weighting and a scoring scale to help determine whether a contractor should be accepted or rejected.

The weighting in this list is only an example of how a subjective metric system can work. The actual criteria will vary according to what information can be obtained and a judgment on the experience of the evaluation team. Single-person evaluations, even with a strong checklist with clear criteria, are judgment calls – and there should be a check on the results that includes an in-depth discussion of the criteria between the two evaluators. The evaluators should perform their evaluation independently and meet with a detached third person for the discussion. The goal is not to determine who is wrong/right/more accurate, but to gain the most objective view for the most objective decision possible. There are no perfect contractors – no organization will score 100. The trade-offs will be centered on the various factors important to the project – price, schedule, function, reliability, operation expense, and operation income.

TABLE 2.1
Contractor Evaluation Weighting Criteria

Criteria	Weight (%)	Scoring Scale (1–5)	Evaluation Notes
Past Performance	**30%**		
Experience Modifier Rate (EMR)	5%	1 = Poor, 5 = Excellent	EMR < 1.0 preferred
Total Recordable Incident Rate (TRIR)	5%	1 = High, 5 = Low	Compared to the industry average
Lost Time Injury Rate (LTIR)	5%	1 = High, 5 = Low	Lower rates preferred
Regulatory Violations (OSHA, EPA)	10%	1 = Multiple, 5 = None	Review OSHA history
Lawsuits and Legal Disputes	5%	1 = Frequent, 5 = None	Ongoing litigation is a red flag
Operational Effectiveness	**25%**		
Schedule Adherence	5%	1 = Frequent delays, 5 = Always on time	Review past projects
Budget Adherence (Change Orders)	5%	1 = Frequent, 5 = Rare	Excessive changes indicate poor planning
Warranty Claims and Rework Rates	5%	1 = Frequent, 5 = None	Indicates workmanship quality
Training and Competency Records	10%	1 = Poor, 5 = Extensive	Evaluate certification and skill development
Systems and Programs	**20%**		
Employee and Contractor Safety Manuals	5%	1 = Outdated, 5 = Updated	Ensure alignment with best practices
Onboarding and Offboarding Programs	5%	1 = Weak, 5 = Strong	Verify structure and execution
Competency Verification Program	5%	1 = Minimal, 5 = Rigorous	Should include practical assessments
Leadership Training Program	5%	1 = None, 5 = Comprehensive	Critical for long-term performance
Safety and Compliance	**25%**		
Safety Inspection Program	5%	1 = Poor, 5 = Strong	Review frequency and depth
Hazard Recognition and CAPA System	5%	1 = Weak, 5 = Proactive	Look for integration into policies
360-Degree Referral Verification	15%	1 = Negative feedback, 5 = Positive	Contact past clients and partners
Bonus	**20%**		
OSHA VPP	15%	5 = VPP Status	Highly developed safety program with oversight
OSHA Sharp or ISO 45001	5%	5 = Sharp status or ISO 45001 registered	Well-developed safety program with maintenance activity

2.14 SCORING GUIDELINES

Total Score Range:

- 85–100: Highly recommended contractor
- 70–84: Acceptable contractor, minor concerns
- 50–69: Caution, requires further evaluation
- Below 50: High-risk, consider rejecting

Use this checklist as a structured method to evaluate potential contractors based on key factors that influence their ability to deliver safe, high-quality, and reliable work.

2.15 FINAL DECISIONS

Dividing potential criteria for contractor evaluation helps ensure a more complete and comprehensive assessment. The negative consequence is creating the illusion that the evaluation is objective and quantifiable. Relationships between the people who will be working together to complete the project are important and likely the main deciding factor among other criteria. If the company does not have the structure to support the project, the best personal relationships are meaningless to project success. Conversely, a highly competent company without constructive working relationships among the people will likely fail. The company must have a long-term mindset that extends beyond the single project to subsequent work with the same project owner or general contractor. Single-project focus will force a short-term strategy that will cause issues for project owners and general contractors with long-term aspirations.

Use the selection criteria to inform a decision. All decisions will have a subjective, emotional component. The criteria can be used to justify an emotional decision. The best way is to create a plan that is as objective as possible and then use instinct and emotional judgment to fill in the inevitable gaps.

2.16 CONCLUSION

Selecting the right contractor requires balancing objective metrics with subjective assessments. While safety data and compliance records provide an initial screening mechanism, deeper evaluations of leadership, competency development, and operational stability are critical. Contractors that prioritize workforce development, foster continuous improvement, and uphold a culture of safety and accountability contribute most effectively to project success. Organizations should use this structured evaluation framework to make informed, strategic contractor selection decisions.

2.17 ARTIFICIAL INTELLIGENCE TOOLS

ChatGPT – with good prompts, can be effective in indexing and searching the web for research data. The results can be verified with the agencies and with the potential contractor.

Injury search prompt: Search for reported worker injuries, safety violations, OSHA citations, or workplace accidents involving [Employer Name] in the United States or Canada. Include sources such as OSHA databases, state/provincial safety agencies, news reports, and legal filings from the past 10 years.

Environmental incidents and violations prompt: Search for environmental incidents, permit records, permit violations, environmental compliance actions, and regulatory enforcement cases involving [Company Name] in the United States and Canada. Include data from the US EPA (ECHO database), state environmental agencies, Environment and Climate Change Canada (ECCC), provincial ministries of environment, and news sources from the past 10 years.

Potential reputational damage prompt: Search for negative public information involving [Company Name] in the United States and Canada, including news articles, legal actions (civil or criminal), employee lawsuits, social media controversies, public protests, regulatory scrutiny, and reputational damage by association with partners, suppliers, or projects. Focus on incidents from the last 10 years and cite credible sources such as court records, news outlets, regulatory filings, and social media trends.

BIBLIOGRAPHY AND FURTHER READING RESOURCES

Helpful Websites

EPA ECHO Facility Search: https://echo.epa.gov/facilities/facility-search (tip for use – this is an all-encompassing database that includes permits and not just violations and enforcement actions. Verify locations and addresses and cross-check with the facility EPA Identification Number for the facility. The number stays with the facility regardless of actual ownership, so additional research is required to determine actual responsibility for any event or permit).

Incidence Rate Calculator and Comparison Tool: https://data.bls.gov/iirc/ (tip for use – this uses the TRIR data from the OSHA 300A form (www.osha.gov/sites/default/files/OSHA-RK-Forms-Package.pdf) and can be useful for a quick calculation and general comparison of TRIR between entities).

NAICS Code Lookup: www.naics.com/search/ (tip for use – use the drill-down and the search to verify the correct code).

NAICS Code Lookup through the US Bureau of Labor Statistics (BLS): www.bls.gov/iag/tgs/iag_index_naics.htm (tip for use – start with "Industries at a Glace Home" and drill down).

OSHA Establishment Search – looks for violation history from OSHA or reported to OSHA for State Plan states: www.osha.gov/ords/imis/establishment.html (tip for use – use multiple variations of name for the contractor and potential DBA (Doing Business As); cross check with location. There are likely multiple entities with the same name so you have to verify with both name and location).

US BLS TRIR and LTIR (Lost Time Incident Rate) data by NAICS code: www.bls.gov/web/osh/table-1-industry-rates-national.htm (tip for use – this data is based on aggregate reporting for larger businesses; 100 or more employees in high-hazard NAICS codes and 250 or more for all employers (www.bls.gov/web/osh/table-1-industry-rates-national.htm). Smaller employers will have significantly larger swings in TRIR and LTIR due to the multiple used in the calculation of the ratios. Additionally, the total number of hours likely includes administrative staff with significantly lower risk profiles).

3 Better Contractors Are Selected When Safety and Procurement Collaborate

Samuel Rayborn

3.1 INTRODUCTION

Global supply chains are complex, and managing vendor relationships is critical. The primary function of procurement is acquiring the necessary goods and services needed for business operations. However, the role of procurement extends beyond the simple act of purchasing.

In today's modern business climate, safety is paramount, as is risk avoidance and/or transference. Those organizations that successfully navigate these turbulent waters do so with an integrated approach in which vendor selection by procurement is achieved in collaboration with compliance and safety as peers in the decision-making process. Procurement, safety, and compliance teams must work in harmony to mitigate risks and ensure quality. Building a world-class team requires strategic planning and an integrated approach.

For the sake of clarity, in this chapter, I will use terms like supplier, vendor, and contractor interchangeably without infusing any additional nuancing into their meaning.

3.2 STRATEGIC STEPS FOR ESTABLISHING COLLABORATIVE PRACTICES

While the following list is by no means exhaustive, it forms a good starting point for establishing collaborative practices among your teams. As you begin going down this journey, you may discover that additional teams need to be included, and additional steps added to the list. That is precisely how this recommended list is supposed to function – as a springboard for more ideas and implementations.

3.2.1 Defining Roles and Responsibilities

Each team must have clear roles. Procurement focuses on vendor selection and contract management, safety oversees vendor adherence to safety standards, and

DOI: 10.1201/9781003679448-3

compliance ensures that all legal and ethical standards are met. Clear role definitions prevent overlap and promote accountability (Smith, 2020).

With regard to contractor assessment and selection in particular, procurement teams play an important role. After all, they are tasked with selecting suppliers and contractors that meet quality, safety, environmental, legal, financial, ethical, and any other risk factors as identified by the host organization. In practice, this evaluation is routinely termed "pre-qualification" and is typically handled by a third-party software as a solution (SaaS) providers, like Veriforce. The rigorous prequalification process ensures that contractual documentation is signed, insurance requirements are met, financial stability is confirmed, and an assessment of the supplier's safety record is performed. Although the depth of the safety assessment may vary, it is the author's experience that the assessment at a minimum includes a review of past incident rates, safety certifications (e.g., ISO 45001), completion of a safety questionnaire to confirm adherence to relevant safety regulations, a review of policies and procedures, and attestation of employee training. More in-depth measures include audits and assessments to verify execution. This lengthy but necessary method ensures that only reputable vendors are selected, reducing the risks of legal issues and accidents over the life of the contract. Additionally, by selecting vendors with strong safety practices, procurement can minimize disruptions due to accidents or regulatory violations, thereby avoiding associated costs.

3.2.2 Implementing a Collaborative Framework

Create a framework that facilitates regular communication between procurement, safety, compliance, and other relevant teams. In many organizations, these departments operate in vacuums and only interact when issues arise, such as an accident, a compliance issue, cost overruns, or other post-hire or onboarding problem of the particular vendor. This late-stage interaction usually results in a finger-pointing exercise that is unnecessary friction within the organization and could have been mitigated prior to the incident or event that has caused the issue in the first place.

Joint meetings and shared databases can bridge gaps, ensuring everyone works toward the same goals. Technology solutions like enterprise resource planning (ERP) systems can integrate processes, enhancing collaboration and data transparency.

3.2.3 Developing Expertise

Invest in training and development. Encourage continuous learning through workshops and seminars. Certifications in safety and compliance can be beneficial. Cross-training among the teams can also enhance understanding and cooperation.

3.2.4 Vendor Selection Criteria

Develop stringent vendor selection criteria that include safety and compliance benchmarks. Use a scorecard approach where vendors are rated based on various parameters, including past performance in safety and regulatory adherence.

As contractors are selected, procurement teams are usually at the heart of negotiating and establishing contracts that include specific clauses related to safety standards and regulatory compliance. Procurement contracts should clearly describe the financial, indemnity, quality, production schedule, and safety requirements. Safety requirements within contracts should not omit such important categories as contractor responsibilities for incident reporting, incident management, investigations, and corrective actions. Additionally, penalties for safety violations should be well-defined.

Establishing a collaborative approach; however, is just the beginning of your journey. Now that you have identified the right teams and taken the first steps to ensure these teams are on the same page and begin to communicate, what comes next? The following section begins to answer this question by providing some data on how successful teams have progressed in their collaborative journeys.

3.3 BEST PRACTICES ON COLLABORATIVE APPROACHES TO VENDOR SELECTION AND MANAGEMENT

3.3.1 Risk Management

Adopt a proactive risk management approach. This includes establishing a risk register to identify and mitigate external and internal risks and considering the entire life cycle of either an individual project or the organization as a whole. Your risk management protocol from the procurement perspective may include using tools such as failure mode and effects analysis (FMEA) and hazard and operability studies (HAZOP) to identify potential risks in safety-sensitive operations.

Conduct regular audits and assessments to identify potential risks. Engage vendors in joint risk management workshops to share insights and develop common solutions (Nelson, 2020).

3.3.2 Continuous Improvement Culture

Encourage a culture of continuous improvement. Regularly review processes and seek feedback from vendors. By facilitating training programs and collaborative efforts, procurement helps vendors improve their safety procedures and compliance with regulations. This investment in vendor capabilities can lead to a more reliable supply chain, reducing delays and potential downtime costs.

Implement changes based on insights gained, aiming for enhanced efficiency and compliance.

3.3.3 Leveraging Technology

Utilize technology to monitor vendor performance in real time. Automated systems can track compliance and safety incidents, providing actionable data that teams can use to make informed decisions. Find ways to involve procurement in regular monitoring and auditing of vendor practices to ensure continued compliance, early risk identification, and prompt implementation of corrective measures.

3.3.4 Strong Relationship Management

Build strong relationships with vendors based on trust and transparency. Regular communication and involving them in strategic planning can enhance cooperation. Clear communication channels help resolve issues swiftly.

3.4 BENEFITS OF COLLABORATION

A collaborative approach to vendor selection and management is beneficial for hiring clients and contractors alike. First, there is the financial impact. A combined focus on contractor compliance enables procurement to play a significant role in helping the host employer to avoid fines, legal fees, and compensation costs associated with accidents and violations. Improved safety and compliance lead to fewer workplace accidents, resulting in lower insurance premiums and related costs.

Second, there is the operational impact. Effective safety practices and compliance ensure smoother operations, reducing downtime and maximizing productivity. A smooth supply chain operation has a direct positive impact on revenue and profit margins.

Third, there is the reputational impact. A strong compliance record can enhance the company's market reputation, attracting more customers and retaining existing ones. This can, in turn, increase sales and market share, positively impacting the company's financial performance.

3.5 CONCLUSION

Building a world-class procurement, safety, and compliance team involves strategic planning and dedication to continuous improvement. Within this plan, a well-integrated procurement team can make a significant impact on vendor safety and regulatory compliance by collaborating with safety and operational teams on the enforcement of the high standards through selection, contracts, and monitoring of contractor performance. Together, these and other teams can protect the financial integrity of the hiring clients, contribute to erecting more integrated and agile practices, enhance vendor experience, and advance the overall improvement of the supply chain.

BIBLIOGRAPHY AND FURTHER READING

Nelson, K. (2020). Proactive Risk Management in Global Supply Chains. *Risk Management Insights*, 30(5), 48–54.

Smith, J. (2020). Defining Roles in Procurement and Safety. *Journal of Business Strategies*, 9(7), 72–78.

4 Contractor Partnership

Ron Gantt

4.1 INTRODUCTION

In a perfect world, contracting would be quite simple. Customers would just clearly tell the contractors what they want, and contractors would make it happen up to that standard and then be compensated reasonably for it. There may be some challenges if conditions differ, and the contractor needs to do things differently than initially agreed. However, these changes would be quickly identified, discussed, and managed appropriately by all parties. No problem.

If you're reading this book though, you know that, in practice, it doesn't really work out that way. Interactions between contractors and clients, or between contractors and other contractors, are often messy, sometimes frustrating, and occasionally catastrophic. Regarding catastrophic interactions, consider, for example, the BP Deepwater Horizon disaster or the shuttle losses from NASA (Challenger and Columbia). In all these cases, difficulties between the contractor and their customer or between different contractors were implicated as issues contributing to the disasters. Subsequent research bears this out: the increased use of contractors and third parties in industries such as industrial construction and transportation is correlated with increased numbers of accidents.

Setting aside disasters and accidents, the day-to-day interactions between contractors and customers can be quite frustrating and lead to other forms of loss, such as wasted time, money, schedule disruptions, quality issues, strained relationships, etc. Anyone who has worked at job sites requiring interaction with a third party – whether it is a contractor or customer – has likely experienced moments where work feels like babysitting, ensuring that the other party doesn't blow things up for you, which, of course, pulls your attention away from your real job.

Unfortunately, not only is this experience common, but there's also no sign that it will improve anytime soon. In today's increasingly global and interconnected world, the use of third parties to get work done is growing. This means that not only is the use of contractors, along with all the challenges that arise in a contractor–client relationship, growing, but this increase will have complex effects that will compound the problems. Take, for example, the industry I work in, data center development. The demand to build data centers is so big that we need a large pool of skilled contractors to work with in order to meet these demands. However, procuring this skilled labor is not as simple as just snapping your fingers and making a busload of licensed journeyman electricians appear. The contractors we use are getting stretched because they are all relying on a similar pool of skilled labor and subcontractors. So, they are pulling in labor wherever they can get it, leading to more and more tiers of subcontracting

 DOI: 10.1201/9781003679448-4

relationships and reduced competency and experience in the workforce. How the industry will adjust to these issues remains to be seen.

Of course, this isn't the contractor's or anyone's fault. This is a symptom of something that researchers call *the multiparty dilemma*. The multiparty dilemma is that, increasingly, to keep up with demand, organizations are forced to rely on third parties. At the same time, increased reliance on third parties creates new challenges and new opportunities for things to go wrong. We've seen this in a variety of industries, from the disasters mentioned previously all the way to the recent major internet outages involving Microsoft and Crowdstrike, which led to significant disruptions in the aviation and healthcare sectors worldwide. So, organizations are increasingly put into a tough position – they don't have the capability to do all the work in-house to keep up with demands, but contracting out creates new problems. To paraphrase U2, we can't live with or without third parties like contractors, customers, consultants, and so on.

This chapter focuses on how we can work with third parties better. We will start with a brief review of some of the research on what makes good coordination in teams, whether it is internal or with third parties. There are consistent factors that must be in place or taken into account to achieve effective teamwork. Then we will talk about how contracting relationships inherently make good coordination difficult by inhibiting or eliminating the necessary factors. Finally, we will give some recommendations on how to make third-party coordination more effective.

4.2 COMMON GROUND AND COORDINATION

One of the most important factors that supports good coordination is common ground. Researchers in coordination and joint activity (i.e., when two or more entities carry out an extended set of behaviors that rely in some way on each other) define common ground as the pertinent set of mutual knowledge, beliefs, and assumptions that support some joint activity. To illustrate what this means, think of how people use the term "common ground" day-to-day. We talk about finding common ground with people, as in, we find that we have things in common in what we believe or expect, which allows us to move forward. Why is common ground necessary to move forward? It is necessary for many reasons, according to researchers, one of the most important of which is this – coordination requires interpredictability, or the ability to predict what the other will do in some circumstances.

Think of an easy example – buying a cup of coffee at the local coffee shop. This typically involves you coordinating with one or more people at the shop, but let's just think of the coordination between you and the person behind the cash register. You walk up, and they greet you. Now think of all the responses you could make that would be acceptable in this situation, such as telling the person what kind of coffee you want, greeting them back, asking them a question about the coffee, asking for more time to decide, and so on.

Now think of all the responses you could make that would be . . . less than acceptable. You could just sit there and stare at them. You could give them a hug or a kiss. You could turn your back on them. You could place them under citizen's arrest. And so on.

Why are the acceptable responses acceptable, and the unacceptable ones not? Common ground. We have in our society an understanding of how the coordination between us and coffee shop workers is and is not supposed to be. We know that, most of the time, our interactions should be brief, involve very little personal exchange, and focus primarily on the purchasing of coffee. There can be deviations from this (which we'll get to), but most of us have this understanding, which allows the coordination between us and the coffee shop worker to proceed smoothly most of the time. We don't have to negotiate how to interact. We don't have to go to team-building exercises and do trust falls with the coffee shop worker. It just happens because we, as a society, have common ground around this joint activity.

Notice how this shared common ground creates efficiency (we don't have to waste time talking about how to do it every time), and it's rather resilient (we go to pretty much any coffee shop with any kind of worker who may have undergone various types of training and may have very different personalities or be in various mood states, and it works pretty much the same way every time). Think of what would happen if one or both of the people involved didn't have this common ground, as in they didn't know how this interaction was supposed to go or what the other person was supposed to do. If you, for example, handed the person some money and they just stared at you wondering if you were giving them a gift. The interaction would be awkward, slow, or just wouldn't work.

That's the importance of common ground. It creates the potential for resilient performance even among people who have never met each other before.

However, one of the big catches with common ground is that it can break down (or not be created at all). People aren't born with common ground. It has to be learned. And if one of the parties doesn't have it or chooses to violate the common ground, the whole thing falls apart.

Think of a situation where you are doing a project with someone. You believe that both of you are aware of the deadline, and so you assume that you share common ground around the need to finish each other's portion before the deadline. But what if the other person is assigned another project that they believe is of higher priority and decides that finishing their portion of your project on time isn't necessary? They may assume that you would agree with them, but if you don't agree or if they don't share that assumption, it may lead to you acting with incorrect information (i.e., a breakdown in common ground).

Because the world is messy and always changing, it creates all kinds of opportunities for common ground to break down almost every day. This means that common ground isn't something you just create and then you're done. Instead, it's dynamic. It can grow or decrease or disappear altogether without warning. This creates a need to monitor common ground, that is, to occasionally check to see what other people are doing and if their current version of the world is the same as your current version of the world.

And this leads to another critical aspect of coordination that we must consider in order to make coordination work well – coordination costs. To understand coordination costs, think of a task that you are assigned that takes one hour to complete, and you need to do it eight times, resulting in a total of eight hours of work. Now imagine you are given another person who is equally competent as you, who will help you do the task. Will it take half the time? It depends. If you two can work on the task independently, as in you can split up the tasks so that they do half and you do half and neither one of your works will really affect how the other works, then it probably will take half the time.

But what if the other person doing some of the task affects how you do the task, where completing your task the right way depends, in some way, on them doing their task at a certain time or in a certain way? Now your task requires coordination, where you will need to keep an eye on each other to make sure you're in sync. That extra effort to monitor the each other's work is an example of coordination costs. It's extra work, effort, resources, or friction that are expended or encountered just because of working with others.

And this creates a challenge where working with others can be either advantageous or disadvantageous, depending on the circumstances. When working with others works well, you become more efficient. But when the escalation of pressures turns your current circumstances into a high-stress and high-demand environment that results in a breakdown of communication and processes, working with other people becomes much more effortful than it might be worth because of the costs of coordination. This is ironic as those same moments when we could most use the benefits of working with others, due to the extra resources they can provide, may cause us to withdraw from the joint activity because the costs of coordination are too high.

4.3 WHAT MAKES COORDINATION CHALLENGING WITH CONTRACTORS

The previous discussion starts to hint at what makes the contracting relationship so difficult. Remember that a key to good coordination is building common ground and keeping an eye out to identify where common ground is breaking down so you can repair it quickly. By their very nature, contracting relationships make building and maintaining common ground difficult. To understand the reason behind this, think about the concept of coordination costs. What are the costs that people in either organization must bear to build and maintain common ground in contracting relationships? Or, to ask this question another way, what makes effective coordination difficult in this situation?

One answer to the question is in the very nature of the contracting relationship. Either for practical reasons (i.e., the contractor has expertise that the client does not have) or legal reasons (i.e., the client wants to keep the contractor at arm's length to avoid dual employment liability issues), there are factors that make it structurally difficult to keep an eye on and intervene in what's happening from one organization to another. A customer supervising the work of a contractor that doesn't have expertise in the work the contractor has may not know enough about the work to identify that the common ground is breaking down until it's too late. Alternatively, some clients avoid monitoring contractors altogether to avoid liability risks, which makes monitoring common ground almost impossible.

Another difficulty related to the nature of the contracting relationship is that contractors are regularly temporary or are kept separate, meaning that they often have little or no ability to develop the context needed to build or maintain common ground. A simple example of this is in contractors knowing how to navigate the nuances of safety requirements from site to site. Imagine a conscientious contractor who genuinely wants to do the right thing and really cares about employee safety. How does that contractor know what's required of them? They use a combination of experience (shared, industry-wide common ground) and whatever information the client gives to them in advance (which is often haphazard in my experience). This means the contractor will try to do the right thing but may make some mistakes, such as filling out the paperwork incorrectly. The client sees this as the contractor not executing the safety program effectively, and the contractor looks subpar. But that's just because both sides thought they shared common ground about what was required, but both were wrong.

Finally, and perhaps most importantly, most contracting relationships have authority differentials, meaning that one side, typically the client, has more authority than the other. This creates all kinds of interesting dynamics, but most importantly, it creates an incentive for the contractor to control the flow of information from them to the client. Anyone reading this who is a contractor is likely nodding their head right now because you know exactly what I'm talking about. Many contractors expend all kinds of extra effort reviewing, wordsmithing, approving, and, in some cases, hiding information that goes to the client. These are costs of coordination that not only add extra work, but also make building and maintaining common ground much more difficult.

But this isn't the contractor's fault necessarily. Clients often encourage this by taking a blame-oriented stance toward contractors or by "rewarding" contractors' transparency with extra work. I've been in situations where we started being transparent to the client about things like safety incidents, and the client started requiring difficult, unnecessary actions from us. And so, we opted to scale back our transparency as a result. Why get punished for doing the right thing?

The bottom line is that the very nature of contracting relationships, perhaps ironically, makes successful coordination very difficult because the costs of building and maintaining common ground are high. As a result, both sides are encouraged to do the minimum and keep each other at arm's length and just deal with the frustration that inevitably occurs.

4.4 SUPPORTING SUCCESSFUL THIRD-PARTY COORDINATION

However, it doesn't have to be this way. One reason we can say this is that most of us have worked on projects where this isn't the way the contracting relationship works. We've been at job sites where people work well together, there is transparency, and people help one another instead of blaming each other. In many of these projects, we often make friendships that we carry on to other projects. So, we know that a better way is possible. But how?

In my experience, the solution needs to start with a mindset shift. A lot of times, people approach contractor coordination as a problem of "how do we get people to do what we need them to do?" The basic idea here is that the problem lies in people not complying with our standards. Now, it may be true that people aren't complying, but the source of the problem in the earlier question is the people themselves. Instead, what we have described in this chapter is a different problem, a problem inherent in the messy system of contracting.

So, the information provided in this chapter can be utilized as the foundation for a better design of the client–contractor relationship. Perhaps the journey to a better relationship begins with asking this question: "How can we design our contracting system to support managing the costs of coordination and build common ground?" To me, this is a more fruitful question because it widens our gaze to the broader system issues and isn't focused on fixing the problem one contractor at a time.

With that as a backdrop, the next place to intervene is in psychological safety at the site. Psychological safety is a bit of a buzzword, and it sounds fluffy, but it's not. Psychological safety means that people can take risks – including pissing each other off – without getting in trouble if it's done in good faith. So, for example, in a psychologically safe environment, contractors should be able to tell clients where they are or aren't making things more difficult, and there shouldn't be repercussions. There should be space for open sharing of information across the boundary. Note that this is risky. Sharing information or relaxing the ability to blame people for mistakes may create risks of getting in trouble, letting people off the hook, etc. But it's important to remember that not having psychological safety is risky as well, because it encourages people to hide things. You can't fix problems you don't understand or even know about. So, for me, I'd rather risk having psychological safety and deal with the risks it brings, because at least then I know about the risks I'm facing rather than not have psychological safety and be flying blind.

Building psychological safety, though, is easier said than done, particularly with contracting relationships because, trust me, contractors have heard it all before about wanting to build partnership and trust and blah blah blah. That's why you can't build psychological safety with words. You only build psychological safety through actions, and the most important thing you can do is to lean in to trust. Show people you trust them, and they are more likely to trust you. What this looks like will depend on your exact situation, but it's worth asking where, in your system, either intentionally or unintentionally, you are not trusting the people on the other side of the contracting relationship, and where you can relax that distrust a bit in the name of building psychological safety. If you do it intentionally – telling people what you're doing and why, and a little bit at a time as a gesture of trust-building – it will often pay off in the long run.

Second, because coordination has costs associated with it, I recommend we designate someone who has the function to manage coordination. This may not be one dedicated full-time person, but maybe someone whose role is at least partly to manage how the sides are working well together. The metaphor I like to use is that this function is like a host. Think of a good host who has people over to their home. That person doesn't dictate how people in their home should have fun. They do things like set the table for people, give them guardrails about places they should or should not go, they make things available to people but give them a choice about what they may want, they keep an eye on people to make sure they are having a good time, and so on. In the same way, a host in a contracting relationship could take the role of looking at all the ways the two (or more) sides need to interact and how to make that go well. Are the meetings in the right places with the right people, at the right times, talking about the right things, with the right information available? This person (or persons) could monitor how we are monitoring common ground breakdowns, keeping an eye not just on whether or not the contractor is achieving their goals, but how the coordination is taking place between the two parties.

This recommendation may seem strange to people, because it's silly to require someone to be there just to make sure everyone gets along. But if you think about it, this happens already most of the time. There are always people at the job site who are trying to build relationships across the boundary with others to keep things moving. There are people who are focusing on how the relationship is working with the contractor. All I'm recommending is to make this an intentional role rather than one people make up on their own. That way, we can dedicate appropriate resources to making it happen rather than doing it haphazardly.

Finally, something that happens at almost every job site is that people try to find ways to build relationships with one another. This is partly just people being friendly, but it's also people finding ways to grease the wheels and help the project. In a sense, you can think of this as people building common ground on a personal level to facilitate building common ground at the project level. This is something that should be encouraged and made more visible. Projects should build in opportunities for people to do this rather than having them do it on their own. Not only will this facilitate sharing information and building common ground, but it often makes projects more enjoyable. And doing it intentionally as an organization also helps make it more transparent, which helps deal with the sometimes-shady things that can happen when relationships are built across boundaries.

4.5 CONCLUSION

Much of this chapter stems from my own research, which I was able to do in my own organization as part of my doctoral studies. In that research, one of the things I found is that not only are contracting relationships extremely difficult, but people find ways to bridge the gaps and keep things running by adapting and coordinating in ways that are often unseen. Put another way, a lot of the costs of coordination in contracting relationships are borne by individuals on projects who go above and beyond their job roles to make things work. This is bad not only because these people often go unappreciated, but also because if organizations don't realize this is happening, they may inadvertently consider these things violations and do something to eliminate these coordination efforts. As a result, they may end up eliminating the source of success for their projects. Anyone who has worked in contracting for a while knows that what I saw in my research is true. People find a way to make things work often without the help of, or in spite of, the organizations involved.

Instead of fighting though, knowing that this is happening, we have an opportunity. By shifting our mindset away from seeing contractors as the problem to fix, we can see the problem as one of a messy, complex contracting system that makes coordination hard. Then we can work to find ways to build and maintain common ground by fostering psychological safety, building relationships across boundaries, and being good hosts with our trade partners. Not only will this give us projects with fewer frustrations, but it will also give us the potential to perform resiliently even with the uncertainty inherent in the contracting world we work in.

5 Client–Contractor Collaboration and Ways to Achieve It

Chris Detillier

5.1 INTRODUCTION

Strong, trusted relationships are the foundation of every successful endeavor. Healthy relationships can provide benefits for our overall well-being, such as providing emotional support, a sense of belonging and joy, personal growth, and mentoring opportunities. Relationships offer support to accomplish goals, fuel success, and offer mental and physical health benefits. Relationships shape how we communicate, how we resolve conflict, and how we navigate challenges that arise when people work toward a common goal.

Nowhere are quality relationships more critical than in multiemployer worksites where clients, multiple companies, and contractors must operate next to each other, often under tight deadlines, high stakes, and hazardous conditions. In these environments, establishing effective relationships between clients and contractors is essential and must be built on trust, transparency, and mutual respect. Collaboration is the key driver of innovation, efficiency, safety, and overall project success. In this chapter, we will explore how organizations can cultivate these relationships.

5.2 CLIENT (PROJECT OWNER) PERSPECTIVE

Ask different hiring clients how they would rate their relationships with contractors, and you will get a host of different views. Clients may perceive their role through a mixture of lenses. Some may approach contractors from a legal and risk-based standpoint, as in how we can de-risk or transfer most risk onto the contractors, whereas other clients may use a strategic and collaborative approach, as in we succeed or fail together, so contractors are our extended family. The complexity of the project, contracts, and the client's organizational culture shape the approach that they follow.

Clients will usually fall into one of the following four categories when it comes to contractor management:

1. Reduced oversight. This is the most basic approach to managing contractors. Clients who utilize this approach take the following stance – we need systems in place to ensure you (contractors) are capable before you arrive. In these situations, the client will aim to streamline oversight and expect

DOI: 10.1201/9781003679448-5

contractors to meet minimum prequalification requirements and to have strong internal controls. Clients will generally tend to focus on strategic decisions, and the operational responsibility will shift more toward the contractor. Often, contract language shifts most of the responsibility for safety, training, and compliance to the contractor. If not managed properly, this approach can go from managing risks to risk exposure. From a practical standpoint, there is little "warm and fuzzy" feelings between the client and the contractor, as their interaction is viewed as strictly business.

2. Balancing cost, schedule, and safety. Clients who utilize this course take the following stance – we expect high standards and compliance, but we also need to make sure that we stay on budget. In many cases, this can create tension with contract companies and crews and may lead them to cut corners or take shortcuts. When clients put a heavy emphasis on cost and schedule adherence, safety will be pushed out of significance either directly or indirectly. Even if the clients do not explicitly communicate that safety is not as important, the message to the contractor to prioritize cost and schedule will be loud and clear. So those clients who want to balance cost, schedule, and safety need to be very explicit about what this approach does and does not mean with regard to their contractors.
3. Site control. Some clients choose to function as site controllers or gatekeepers. Their perspective related to contractor management is quite simple – we expect compliance and quality work. If the contractor does not deliver, then we will find someone who will. This method is more of a hierarchical approach – the client sets rules, and contractors are expected to comply out of fear of losing work. They are monitored closely, with heavy oversight, and trust is earned through proven performance. Though this method may work by replacing low-performing contractors and holding them accountable, it can also lead to low morale, lack of trust, limited communication and collaboration, and discourage reporting out of fear. Unfortunately, this approach can suppress feedback and eliminate any collaboration from contractors.
4. Coordination and direct control over the work and safety of contractors. Some clients choose to act as coordinators and take direct control over the work and safety of contractors. These clients feel that they need contractors who understand the risks and take ownership of safety and compliance, because ultimately, they feel as though they (the client) carry the reputational and legal consequences. This group of clients is usually involved in the operational side of the work taking place because it assumes the responsibility for setting expectations and maintaining control over safety, quality, and timelines. They expect contractors to adhere to site safety policies, regulatory requirements, and performance expectations. Though this approach provides great oversight and guidance, from a liability standpoint, clients must be aware that they may be held responsible for incidents on their site even if caused by a subcontractor.

While the approaches mentioned earlier may differ in philosophical backgrounds, methodology, and practical outcomes, they do illustrate one simple principle, that, no matter what approach the client uses, their role is more than just awarding contracts. Clients have to balance oversight and the acceptance of risk, while ensuring that contractors understand and follow safe work practices in order to prevent incidents on their assets.

As a client, what approach are you using? Is it one of the four mentioned earlier? Or is it a combination of them? Or perhaps something entirely different? While settling on the right approach may be important, what is even more critical is ensuring that you keep open lines of communication with your contractors, so that they can feel safe to provide open feedback on your sites and leave in a better condition than when they arrived. Are you confident that your current contractor management strategy accomplishes this?

5.3 CONTRACTOR PERSPECTIVE

The vast majority of contract companies will tell you that their role is to execute the contracted work in accordance with the client's expectations and make sure the client is satisfied, in order to win future contracts. This is not always the case, and there is much more to it than just that, but for a substantial number of contractors, this is their reality. In many cases, this reality is formulated by the real-world experiences that the contractor has faced, which in turn shaped their perspectives. For example, when contractors perform service for a gatekeeping, site-control client, they know what is expected of them, namely, getting the job done without much splash. Of course, in those cases where contractors have had positive client relationships or have a higher level of cultural maturity, they may be more inclined to be proactive and speak up when there are concerns or areas of improvement.

The role of the contractor can be complex and requires balance. Contractors have several responsibilities while performing work for clients, which include:

- delivering the work safely;
- adhering to the client's scheduling milestones;
- staying within budget requirements;
- maintaining professionalism;
- nurturing open and proactive communication;
- protecting the client's interests;
- managing ever-shifting supply chain reality;
- and being able to adapt to rapidly changing markets, regulatory parameters, and financial pressures.

This list is by no means exhaustive. However, it vividly illustrates that meeting all the often-competing requirements and expectations almost takes a dance – a thoughtful and ever-evolving give and take between contractors and their hiring clients, built on the solid foundation of partnership and cooperation from both parties. This is how it should be in principle.

In reality, however, many of the contractors that I interacted with do not feel like their hiring clients are willing "to dance." The hard truth is that clients often operate from the perspective that expects only contractors to give. It is definitely not a fifty-fifty scenario.

Contractors understand that contractor management is necessary, but they sometimes feel that the relationship is unbalanced, and they will have to walk a tight line to keep their clients happy. To the contractor, the client holds the power. Many contractors reveal that their input to the client is all too often valued as long as it is convenient, but it is easily and quickly dismissed if it causes any ripples or necessitates changes on the part of the client. Contractors feel as though they are required to be flexible, but the same expectation is not reciprocated from the client side.

When it comes to mutual trust between clients and contractors, many of the contractors do not feel like it goes both ways. Another way of putting it is this – contractors often feel that when anything ever goes wrong, they are the first to take the blame. There is no "innocent until proven guilty" assumption in their interactions with hiring clients. Quite the opposite. They are automatically guilty until they can prove their innocence. And even then, if egos get hurt in the process, the contractors may never regain their "innocence" in the eyes of the hiring client or be hired again. It must be mentioned here that the current generational distribution of the modern workplace greatly contributes to this shift in perspective, as younger managers tend to get offended more easily.

Other contractors feel like they are scrutinized, placed under excessive oversight, or face inconsistent enforcement until they prove themselves to be "trustworthy" in the eyes of the clients, which in and of itself tends to be a very subjective criterion. This can be especially true with new clients. Contractors feel that if they make one mistake, they will be penalized; however, when the client or more established contractors make the same mistake, more lenience is given. This can lead to contractor employees feeling uneasy about the work and scared to make a mistake, which can result in under-reporting, distraction, and psychological safety hazards.

Interviewing multiple contractor companies, especially the field-worker level of their workforce, revealed that the common perception is feeling undervalued, unappreciated, and unheard by the client representatives. Let's face it, in most cases, the contractors know how to do the job and how things work on the ground level. Yet, they are often not brought in until most decisions are made and end up having little if any involvement in early discussions, processes, and planning.

This often results in work inefficiencies, rework, and delays, which could have been prevented if contractors were involved earlier in these processes. This leaves many contractors feeling like their specialized knowledge and experience are underutilized and that they are not part of the team.

Contractors often face pressures, such as budget constraints, tight deadlines, change proposals, and client-specific requirements, etc., that are added by the client. In many cases, contractors experience these pressures being added to them, but feel that they are seldom offered many resources or support by the hiring client. This can lead to cutting corners, haste in getting jobs completed, additional work, and poor morale on jobsites.

Of course, not all contractors have unfavorable experiences with their hiring clients. There are many clients out there that take a partnership approach to contractor management and reap great benefits in return. Likewise, contractors understand that all clients are not equal, and relationships will be different.

Some clients will consider contractors as replaceable vendors, but others use a collaborative, respectful, and team-oriented approach that will earn the contractor's respect. When this happens, contractors will go above and beyond for those clients. Those relationships are built to last and will usually lead to long-term loyalty and prioritization of work. These contractors will see themselves not as outsiders but part of the internal client team. They will be open to communication, provide feedback, strive to carry out the client's vision, and will perform their best work. They will strive to perform well, not out of fear, but out of mutual respect and values.

5.4 IDEAL RELATIONSHIP

The ideal client–contractor relationship is a collaborative partnership. Clients who feel that they are strategic partners with their contractors have a team mindset. They feel that they are in it together, and if they help their contractors succeed, they will, in turn, succeed as well. These clients view their contractors as partners, and not just service providers. The focus is on relationship building and trust through collaboration, open communication, mutual respect, and early and consistent engagement with their contractor network. This leads to the contractor feeling a level of ownership in projects and within the client's organization.

This approach is beneficial because it allows both parties to align project goals, continuously learn from each other, and create innovative strategies. This theme is the ideal approach and can lead to high performance, strong internal accountability, mutual respect, and positive outcomes. It is in these environments where safety management can go from reactive to interdependent, creating more positive cultures for both parties involved.

5.5 BEST PATH TO CLIENT–CONTRACTOR PARTNERSHIP

Positive client–contractor relationships are accomplished through several means. First, clients must set clear expectations for their contractors. How can we provide feedback or hold contractors accountable if clear expectations are not communicated? Expectations can be set through clear contracts and clarifying conversations before and after contract signing. These contracts should specify work expectations, establish minimum pre-qualification criteria that must be met by the contractor to be awarded work, define the roles and responsibilities of both the client and the contractor, set out liability clauses, and describe the actions to be taken in the event of a breach of the contract or failure to meet expectations.

Bridging documents may also be used to communicate and document which health, safety, and environmental policies will be followed when performing work. For instance, if as a contractor, I specialize in crafts such as scaffolding, my scaffolding policy may be more robust than the client's policy on scaffolding. On the other hand, if I am working on a client's site, their emergency action plan (EAP) would be clearer and more concise when on their assets, thus making their EAP plan more applicable. The bridging document would communicate which programs should be followed when working on client sites, so as to eliminate all confusion. Clear contracts and documentation set the framework of expectations between the client and the contract company.

As most clients have contracts in place, those expectations must be communicated down to the crews that are performing the work. In order to accomplish these methods, orientation training and pre-job or job planning meetings can be held with contractors to communicate expectations and plan work. During these meetings and training sessions, clients should set expectations for not only the overall work and safe practices but also teamwork and open communication with the contractors. These discussions should provide opportunities for feedback, dialogue, and be used as an opportunity to socialize lessons learned from past projects.

In order to establish good relationships and also get a sense of the actual company culture, onsite engagement is vital. Many, if not most, of the HSE professionals from both the client and contractor sides that I come across agree on one fact – their need to go out to the field more. So, my advice to all stakeholders involved in the management of contractors would be to spend adequate time in the field.

If you feel like you don't have the time to go out in the field, make the time. It's the single most important investment that you can make in contractor risk management. Being in the field where you can observe work with your own eyes and perform site inspections or audits is beneficial for building relationships, earning trust, and ensuring that work is carried out in accordance with expectations. Visiting the field is necessary to see if policies are being implemented and to address hazards.

When you are out in the field, make sure that your demeanor and approach are not those of the safety police. Instead, use the opportunity to rub shoulders with the contractor workers as the platform for coaching, mentoring, and learning whether there are any gaps between work as planned and work as executed. Of course, any deficiencies found must be noted and corrected, but how it is approached can make all the difference in the world to the contractors and their employees. The overall attitude needs to be one of "we, the assessors, are not out to get you, but to learn and grow together."

As mentioned earlier, and worth reinstating for additional emphasis and clarity, the worksite field inspections and audits are important, but often it is even more valuable to just go where the work is done and build relationships. Get to know the people working on site and show face. Notice their good work, safe actions, and you will not be viewed as someone who only shows up on location when something is wrong, someone is hurt, or shortcomings are noted.

Utilize time in the field to get feedback. Use the following questions and others like them to collect valuable information and build trust with your trade partners:

- Are there any obstacles that are making it more difficult to perform the job? Any safety policies, steps, or protocols that are viewed as more cumbersome than helpful?
- What can we do differently to help you improve your outcomes and working conditions in the future?
- Which steps or parts of the SOPs and protocols do not make sense? Which steps are missing?
- What requirements slow down your progress?
- How would you change or adjust your work processes if you had the chance to do so, in order to make your work more intuitive?

Asking these questions, listening to their responses, and clearly actioning their feedback shows that you consider them important stakeholders and partners in the work, are willing to actively learn from their expertise, and value their opinions. People want to work for someone who cares for them and their well-being. Not only that, but people will also go the extra mile in an environment that makes them feel valued.

Effective feedback is vital for continual improvement and long-term relationships. If there are shortcomings, be sure to address them. We cannot fix something if we do not know that it is broken. Provide timely and effective feedback when expectations are not met.

As mentioned earlier, both contractors and clients alike need to reflect internally as well to see if there are barriers that are being created that could be a root cause of the deficiencies in their relationships. Be open to feedback and work to resolve the issues. There may be times when expectations may continue to not be met, and action may need to be taken, but in most cases, coaching and support can lead to stronger relationships and adherence to expectations.

Remember this too: positive feedback can go a long way, and acknowledging excellent work and celebrating wins and accomplishments can be exponentially effective in creating a culture of quality work and safety performance. When workers are wearing their PPE correctly and dressed in a professional manner, praise those workers for it. Reward crews that always go beyond expectations. This does not have to be anything big, but coin awards, merchandise, cookouts, or shout-outs on company documentation or newsletters make contractor crews feel that they are part of the team and that they are appreciated.

5.6 CLOSING

In conclusion, the client–contractor relationship is not just about managing a project or delegating work – it is about building strong and collaborative relationships. Clients must understand and be familiar with their contractor base. They must communicate clear expectations for their contractors and provide necessary feedback – both negative and positive. They should make contractors feel empowered to voice concerns and provide feedback for continuous improvement.

When contractors feel that they are part of the team and the client wants the best for them through mutual respect, they will go above and beyond expectations. They will want to please the client and perform quality work rather than just ensuring that they do what is necessary to not be reprimanded. Collaboration will lead to long-lasting, positive relationships and will drive risk management from external motivation (I have to) to internal motivation (I want to), where safety is transformed from compliance to value. This is the ultimate goal of a healthy client–contractor relationship. And this is the ultimate goal of safer cultures.

6 Continuous Improvement Requires Employee and Contractor Involvement

Josh Ortega

6.1 INTRODUCTION

Achieving continuous improvement in contractor management depends on the active involvement of both employees and contractors. Their combined engagement fosters innovation, enhances safety outcomes, and aligns with leading global standards like ISO 45001 (2018), ANSI/ASSP Z10 (2019), API RP 1173, and IOGP 423. These frameworks recognize that sustainable safety performance is best achieved when those closest to the work are directly involved in shaping how it is performed and improved.

Too often, contractor engagement in safety programs is passive. In one of our projects, a large subcontracting team consistently participated in toolbox talks and pre-task briefings, but rarely contributed to safety planning. It wasn't until we invited them to co-lead risk assessments and jointly develop control measures that their engagement transformed from procedural to proactive. This shift highlighted how involvement needs to be designed – not assumed. True progress occurs when contractors are not just included in safety discussions, but actively contribute to safety solutions. Effective involvement transforms compliance-driven systems into responsive, learning-centered organizations.

6.2 FOUNDATIONS OF ENGAGEMENT: FROM PARTICIPATION TO CO-DESIGN

Employee and contractor involvement should go beyond surface-level consultation. ISO 45001:2018 defines meaningful participation as direct involvement in identifying hazards, making decisions, and reviewing systems. Similarly, ANSI/ASSP Z10 and API RP 1173 call for workers' voices to be integrated into broader organizational safety thinking.

In many organizations, involvement is reduced to routine meetings or signing off on forms. This approach, often mistaken for engagement, limits the potential for real improvement. When workers do not help create risk controls, they are unlikely to take full ownership of them.

 DOI: 10.1201/9781003679448-6

Genuine participation means co-design, where contractors and employees collaboratively shape policies, procedures, and risk-control strategies. IOGP 423 emphasizes joint accountability and shared oversight as essential for managing risk throughout the contract lifecycle (IOGP, 2016). This shift is also supported by Edgar Schein's model of organizational culture, which emphasizes that true change occurs only when deeper, shared assumptions and values are addressed, rather than just surface-level practices (Schein, 2010).

6.3 CONTINUOUS IMPROVEMENT AND THE DEMING CYCLE

No discussion of improvement is complete without acknowledging W. Edwards Deming, whose Plan-Do-Check-Act (PDCA) cycle remains the gold standard for iterative improvement (Deming, 1986). Deming's framework, widely adopted in safety and quality systems, provides structure to the feedback loop needed to evolve practices over time.

Plan: Define safety goals, identify risks, and develop interventions collaboratively.
Do: Implement those plans in the field, ensuring contractors and employees understand and execute them.
Check: Assess performance using both data (e.g., observations, near misses) and narrative (e.g., worker feedback).
Act: Refine the process, institutionalize success, or start again.

The true test of a safety system is not in how well it prevents error, but how well it adapts when the unexpected occurs. Many contractor management systems claim to be PDCA-driven, yet treat contractors as data points, not agents. Continuous improvement only works if contractors are embedded within each step of the cycle, not observed from afar.

6.4 MECHANISMS FOR INVOLVEMENT

Involvement must be structured into daily operations to be effective. The following list presents some common and effective methods, though many more could be included here:

- **Joint Safety Committees** – where representatives from both the workforce and contractors address safety challenges together.
- **Collaborative Risk Assessments and Job Safety Analyses (JSAs)** – reinforcing shared accountability and creating mutual understanding of work conditions.
- **Learning Teams** – which promote a no-blame environment and draw insights from workers closest to the task (Dekker, 2014).
- **Feedback Systems** – such as digital forms or anonymous suggestions, allowing contractors to raise concerns safely.
- **Cross-Contractor Forums** – where contractors can learn from each other and share practical safety innovations.

These mechanisms show that all contributors, not only employees, have valuable insights that can improve workplace safety. Organizational learning only occurs when systems are designed to genuinely listen and respond.

6.5 BEHAVIORAL EMPOWERMENT AND SAFETY EQUITY

Empowering contractors involves more than permitting them to stop unsafe work. It means transferring real influence over decisions that affect their safety. Scott Geller's Actively Caring for People (AC4P) model highlights the need for positive reinforcement and respectful inclusion in building a strong safety culture (Geller, 2001).

Some organizations, inspired by high-reliability organization (HRO) principles, are testing distributed authority models. These allow the people with the most situational awareness, often contractors performing the work, to make or challenge safety decisions in real time.

Those most exposed to risk should have the strongest voice in how that risk is managed. This shift from hierarchical to risk-based authority does not undermine control; instead, it builds equity into safety governance. As Andrew Hopkins (2008) notes, organizations that rely solely on formal authority and oversight often overlook the insights and intuitions of those actually performing the work.

6.6 CONTRACTOR-LED LEARNING TEAMS: REVERSING THE LENS

Post-incident reviews are most insightful when led by those closest to the work. Contractor-led learning teams offer a powerful opportunity to understand incidents from the field perspective. Instead of relying on internal staff to analyze external contractor events, organizations can flip the model and invite contractors to guide the learning process.

This approach reinforces accountability, builds trust, and uncovers systemic issues that traditional investigations may miss. It also exemplifies a mature safety culture that values humility, reflection, and shared learning.

6.7 FROM SAFETY-I TO SAFETY-II: REDEFINING SUCCESS THROUGH CONTRACTOR INSIGHT

Erik Hollnagel's distinction between Safety-I and Safety-II provides a valuable lens for understanding how contractor involvement enhances system resilience. Safety-I focuses on preventing failure and minimizing harm by analyzing what goes wrong. In contrast, Safety-II emphasizes learning from everyday successes and improving the system's capacity to function well under variable conditions (Hollnagel, 2014).

This shift in perspective has important implications for involving contractors. Contractors often work in the most variable and high-risk conditions. Their ability to adapt in the field, resolve unexpected issues, and maintain safe performance often goes undocumented in traditional safety systems. Safety-II argues that these adaptive behaviors should be studied and reinforced, not overlooked.

Contractor involvement under a Safety-II lens becomes a proactive asset. On one project involving remote field maintenance, our team noticed a recurring deviation from planned procedures by a contractor crew. Rather than penalize them, we invited a conversation to understand why. It turned out their solution addressed a visibility and communication issue during crane operations that had not been adequately considered in the original planning. That discussion led to a formal change in our lift planning templates across the business unit. Learning from these contractor adaptations showed us that real-time innovations are often hidden in plain sight – and deserve our attention. When organizations listen to how contractors make the system work despite uncertainty, they gain insight into operational resilience. Instead of focusing only on failures, they begin to identify the capabilities that lead to success and can strengthen those patterns across teams.

This also requires changing how learning is structured. Rather than waiting for an incident, organizations can conduct reviews of "normal work" with contractors, exploring what went right and why. These reviews illuminate effective practices that can be shared more broadly and built into the system. Ron Gantt goes into more detail on the benefits of proactive learning in his chapter of this book.

In this view, contractor feedback is not supplemental, it is essential because it reveals the human flexibility and system interactions that are too complex to capture in static procedures. By valuing what works, organizations become better prepared for what could go wrong.

Hollnagel's work reinforces the idea that involving contractors is not just about compliance. It is about building a workforce-centered view of safety that supports continuous adaptation and learning. Integrating both Safety-I and Safety-II principles allows organizations to learn from both failure and success and to treat contractor knowledge as a strategic asset.

6.8 TECHNOLOGY AND INNOVATION IN ENGAGEMENT

Digital tools have expanded how organizations engage their contractor workforce. Mobile apps and digital permits have improved access to safety procedures, training, and reporting. More recently, technologies such as natural language processing (NLP) and artificial intelligence (AI) are helping organizations make sense of large volumes of feedback data.

For example, some systems now analyze daily reports to detect patterns in language that may indicate fatigue, confusion, or stress. These early signals allow managers to intervene before problems escalate.

A drop in contractor observations or hazard reports is not just a data anomaly. It may signal disengagement, which requires prompt attention. Predictive analytics can identify these trends and support proactive responses. Similarly, wearable technology is offering new ways to monitor fatigue and exposure. However, these tools are most effective when implemented within a culture that values trust and collaboration.

As Todd Conklin has emphasized, "blame fixes nothing." A system that captures contractor feedback must also be designed to respond to it constructively, supporting learning, not punishment (Conklin, 2012).

6.9 CONCLUSION: TOWARD A CULTURE OF SHARED OWNERSHIP

Engaging internal employees and contractors in continuous improvement is not optional in today's high-risk industries; It is essential. Effective engagement goes beyond issuing policies or completing checklists. It requires building relationships grounded in respect, transparency, and shared responsibility.

Organizations that succeed in this area do more than include contractors. They empower them to lead, contribute, and improve the systems they work within. Achieving this shift from symbolic inclusion to meaningful collaboration involves:

- Designing systems where contractors cocreate safety procedures.
- Supporting learning teams led by those closest to the work.
- Aligning decision-making authority with exposure to risk.
- Using technology to strengthen, not replace, human connection.
- Fostering a safety culture built on mutual trust and respect.

This approach represents the future of contractor management. Rather than treating contractors as outsiders, successful organizations create environments where safety is integrated, equitable, and collectively owned.

BIBLIOGRAPHY AND FURTHER READING RESOURCES

American Petroleum Institute. (2015). *API Recommended Practice 1173: Pipeline Safety Management Systems*. www.api.org

American Society of Safety Professionals. (2019). *ANSI/ASSP Z10.0–2019: Occupational Health and Safety Management Systems*. www.assp.org

Conklin, T. (2012). *Pre-Accident Investigations: An Introduction to Organizational Safety*. CRC Press.

Dekker, S. (2014). *The Field Guide to Understanding 'Human Error'* (3rd ed.). CRC Press.

Deming, W. E. (1986). *Out of the Crisis*. MIT Press.

Geller, E. S. (2001). *The Psychology of Safety Handbook*. CRC Press.

Hollnagel, E. (2014). *Safety-I and Safety-II: The Past and Future of Safety Management*. CRC Press.

Hopkins, A. (2008). *Failure to Learn: The BP Texas City Refinery Disaster*. CCH Australia Limited.

International Association of Oil & Gas Producers. (2016). *IOGP Report 423: HSE Management – Guidelines for Working together in a Contract Environment*. www.iogp.org

International Organization for Standardization. (2018). *ISO 45001:2018 Occupational Health and Safety Management Systems – Requirements with Guidance for Use*. www.iso.org

Schein, E. H. (2010). *Organizational Culture and Leadership* (4th ed.). Jossey-Bass.

7 Charting a Just Course
Where Fairness Meets Accountability

Simon Goncharenko

Creating a climate in which disclosure is possible and acceptable is the organization's responsibility

Sidney Dekker

7.1 INTRODUCTION

In the late 1980s, William Glen Stewart, who had first flown a Tiger Moth as a 19-year-old RAF pilot, was one of the most experienced aviators in the British Airways fleet, with 14,750 hours during over 25 years of service. On November 21, 1989, Stewart successfully – and, some would say, heroically – leveraged his decades of technical expertise and experience to overcome a number of unexpected circumstances while piloting a Boeing 747 on a flight from Bahrain to London Heathrow, in what would later be referred to as the November Oscar case. He overcame the copilot's incapacitation by food poisoning, redirection caused by the weather, clearance from ground control that came too late, zero visibility, and equipment malfunction to land safely, preserving the lives of the 263 passengers on board. Yet, due to the challenges listed earlier, Captain Stewart ended up flying, without realizing it at the moment, at an altitude of less than 80 feet for a distance of 750 feet over a built-up area crowded with rush-hour traffic as he approached Heathrow Airport. Despite the fact that his capable decisions were responsible for preserving all 263 lives on board his aircraft, even before Captain Stewart parked his plane at the gate, phones at the British Airways headquarters were ringing. His accusers held that Stewart blew an instrument approach that he never should have attempted, risking the lives of passengers, as well as numerous motorists and perhaps even hotel occupants.

The UK's Civil Aviation Authority brought a prosecution against the commander, Captain Glen Stewart, under what are now Articles 55 and 56 of the Air Navigation Order 1995, and he was convicted on one charge and acquitted on the other. On May 8, 1991, a jury at Her Majesty's Crown Court in Isleworth, England, hesitantly agreed. In a split verdict (10 to 2), they found William Glen Stewart guilty of negligently endangering his aircraft and passengers. It was the first time in the history of British aviation that an airline pilot was found to be a criminal or was, in fact, even charged with being a criminal as a result of the pursuit of his duties to what he believed to be the best of his abilities.

DOI: 10.1201/9781003679448-7

Taking this matter close to heart, exactly 3 years and 9 days following the incident, Captain Glen Stewart left his small house in Wokingham without a word to his wife. He drove approximately 9 hours to a beach, 10 miles from his birthplace in Scotland, where he attached a hose to the exhaust pipe, led it into the car through a nearly closed window, and in moments, had asphyxiated himself.

In the absence of Just Culture, complex situations are often blamed on a single individual or entity, with nothing good or constructive ever coming out of it. Even at the time of the November Oscar case, the prosecution was controversial and raised serious questions about the policy underlying the decision to institute it. Pilots' Association spokesman Captain Mike Wallis said the decision would tell pilots, flight engineers, and those involved in the industry that, in the future, if they spoke up, they could find themselves facing criminal prosecution. They are going to hesitate before they do it.

7.2 SO, WHAT IS JUST CULTURE?

While many various definitions have been offered to date, in its simplest form, Just Culture is an environment where individuals are encouraged to report errors and safety concerns without fear of blame or punishment, while also holding individuals accountable for their choices and actions. There are two main elements to Just Culture:

1. freedom of reporting errors and safety concerns, including self-reporting, without fear of blame or punishments;
2. and presence of accountability for choices and actions.

To put it differently, think of Just Culture as an environment where we don't react to an incident by immediately blaming the last person at the end of the chain of events that led to it. We investigate. We provide a level of anonymity in the reporting channels and structures for the near misses and what some may consider unsafe actions.

The presence of Just Culture affords us the luxury of learning and improving as a result of the constant inflow of data from the unafraid workers. We can openly identify and examine the weaknesses of our business. Our culture, design, leadership, operational reliability, and systems are investigated thoroughly and corrected not only before but also after errors occur.

7.3 KEY CHARACTERISTICS OF JUST CULTURE

Contractors with Just Culture encourage open and honest communication about errors and near misses. They foster this openness by not blaming their front-line employees for errors because they understand that to err is human. Or, to put it differently, we, the humans, all commit errors: every hour, every day.

Contractors with Just Culture believe in shared accountability, where both the organization and the individual employees within that organization play a role. The organization is responsible for creating safe systems and providing a place of employment that is free from known hazards that could cause severe injury or death.

The individual employees within the organization are responsible for their choices, particularly those characterized by gross negligence, willful disregard, and destructive actions that go against corporate policies and procedures, especially when such actions may directly endanger the lives of others.

Contractors with Just Culture choose to focus on learning instead of blame. Because the environment fostered by Just Culture supports a continual inflow of data from the field, there is always a process that can be improved, a control that can be implemented, or a hazard that can be eliminated. And the learning never stops.

7.4 WHY IS JUST CULTURE IMPORTANT?

Just Culture truly does serve as the contractor's compass, helping them to navigate their companies toward a safer and more sustainable organizational climate. How does it work? It's like this. If you have a Just Culture, your people will not be afraid to report near misses and own up to even their own mistakes; thus, you will foster a Reporting Culture.

By the way, while we are on the subject, your near misses are one of the most important learning opportunities, showing where controls prevented an incident from happening, even in the face of an error. So, if you are not capturing your near-miss data and learning from it as much as you can, you should be.

A Reporting Culture will enable you to have a free flow of data from which you can continually learn, thus fostering a Learning Culture. A Learning Culture will afford you the luxury of knowing what is going well, what is not, what needs to be changed, what needs to be stopped, and what needs to be adjusted. A Learning Culture gives birth to a Flexible Culture. If something isn't going right, you can adjust your actions, your processes, your protocols, and your policies. If there is no learning, you don't even know that you need to adjust, or where to tweak something, or when to do it.

A Flexible Culture allows you to be informed and leads you to have an Informed Culture. You are gathering all sorts of information, and you are constantly putting that information to work by discerning trends, weak spots, error-likely environments, and so on. Collecting and processing the information as it comes enables you to be proactive, as opposed to constantly reacting to unwanted events. Once you have erected reporting, learning, flexible, and informed culture on the foundation of the Just Culture, you are approximating what some may refer to as Safety Culture.

But don't, for a moment, allow yourself to be duped into thinking that once you arrive here, it will remain this way forever. There is no such thing as arriving at Safety Culture. There is only fighting for it and doing everything you can to keep fortifying it on a daily basis, because Safety Culture is really a journey and not a destination.

7.5 ESTABLISHING JUST CULTURE

Establishing the right culture takes time. It will not happen overnight, but the benefits far outweigh the efforts. While a formula for building Just Culture may seem overly simplistic, there are some basic prerequisites without which no success can be achieved. These may be considered required building blocks necessary to sustain Just Culture.

First, people need to be trained on the importance of proper reporting, the required elements of an effective report, and the tools to use in making such reports. Second, your employees need to be provided the right tools for reporting hazards, near misses, and incidents, tools that are easy to use and intuitive. Third, your workers will want to see that their reports are being used, socialized, and leveraged in your organizational learning.

Fourth, establishing clear and effective means of learning from incidents, including investigations, gap analyses, and closure of findings, will go a long way toward establishing Just Culture. Fifth, establishing the environment where constructive feedback and coaching are freely provided and cherished is going to be one of the most essential elements in establishing Just Culture.

7.6 JUST CULTURE TOOLS

Someone correctly pointed out that how your company responds to human failure events can create or destroy a Just Culture. So, one of the best practical tools that I could share with you comes from Todd Conklin's fourth book *The 5 Principles of Human Performance*, published in 2019. The tool is a list of questions that could be asked, or really that should be asked, following an incident:

1. Are the people OK?
2. Is the facility safe, secure, and stable?
3. Tell me the story of what happened?
4. What could've happened?
5. What factors led up to this event?
6. What worked well? What did not work?
7. Where else could this happen?
8. What else do I need to know about this event?

Your reaction as a manager can either keep building on or immediately destroy years of progress in creating Just Culture. So, react wisely and with purpose. Keep fortifying your Just Culture; it will deliver a tremendous return on your investment.

BIBLIOGRAPHY

Conklin, Todd. (2019). *The 5 Principles of Human Performance: A Contemporary Update of the Building Blocks of Human Performance for the New View of Safety*. Santa Fe, NM: PreAccident Media.

Goncharenko, Simon. (2024). *Save Lives: Pushing Boundaries in Human Factors*. Midway, TX: Kindle Press.

Wilkinson, Stephan. (2024). *The November Oscar Incident. Human Factors and Systems Safety*. www.humanfactors.lth.se/fileadmin/lusa/Johan/June_LL_2024/NovemberOscarCLN.pdf

8 One Question
A Window into Contractor's Maturity

Paul Espenan

8.1 INTRODUCTION

I've been working with contractors to understand their challenges and assist them in their safety efforts for many years. These interactions have occurred in many different formats, ranging from a simple phone call, an in-depth meeting, a jobsite discussion, participation in their safety meeting, a lunch meeting, a visit to their worksite, or even a detailed safety validation exercise. These interactions have always been about relationships and understanding context. I've never believed that I was in any position to help, until I could say I fully understood their challenges. It seems that there has rarely been enough time during these brief interactions to fully understand their challenges and have strategic conversations about how we can work together to improve safety.

Over many years, I have witnessed much through the eyes of an operating energy company. I've seen and heard things from contractors I'd like to forget. I have also seen and heard things I will never forget. I've worked with safety professionals who were admired and recognized as leaders across the industry. And on the other side of the spectrum, I've worked with an office administrator for one of our contractors who had no formal safety training or experience. Regardless of the background, mindset, or knowledge, I've always been able to gain new context and build a positive relationship in each situation and every circumstance.

It has been many years, but one story about an office administrator made an impression on me that I will never forget. We were in the process of validating our contractor base and scores in our contractor management system. There were many screening questions that we were asking of our contractors to ascertain their culture and understanding, and practice of safe work.

One particular contractor was experiencing challenges, and their scores reflected a failing grade in our system. The responsibility of figuring out why this company was failing and how to fix their issues fell on the shoulders of their office administrator. Due to the size of this contractor, we were working with their operational leadership, as they were too small to employ a full-time safety professional who could help them decode the safety questions on our questionnaire. So, we started from square one in order to help this uninitiated person understand and improve. We provided instructions, directions, and advice. We even provided sample materials that she could use

DOI: 10.1201/9781003679448-8

to customize their programs to satisfy our requirements. This administrator was a quick learner. We managed to help her supplement her employer's contractor program. The rest was up to her.

Almost a year later the office administrator called me. She reminded me about all the work we did to help her improve. She was very excited and said, "I had to call you to tell you what happened." She said

> we were audited by a large independent energy company that gives us more than half of our work. We passed the audit. If you hadn't helped us out, we would have failed, and a lot of people would have lost their jobs.

She was beyond grateful. All we did was help her get better. We didn't think much of it at the time, but it made a huge difference in her life and in the lives of all her coworkers. This experience changed my whole approach to contractor safety and was a defining moment for me. In that moment, I realized that a short amount of time spent in a conversation with a service company could have a profound impact. My prior ways of focusing mostly on the checklist were over. There was a better way. It would lead to a different conversation.

There are plenty of safety checklists filled with questions, the intent of which is to help us assess our contractors, their programs, their implementation, and their culture. We need checklists to some extent not only because we want to stay organized and be complete in our assessments but also because we are trying to understand how the company is progressing in its safety journey. The challenge of using safety checklists to assess a service company's safety programs is that the "test" can be taught. The "preferred answer" can become the answer, not the actual accurate answer.

Safety consultants can play an important role in improving safety performance. Sadly, I have seen "cut-and-paste" safety programs created by consultants too many times. There is a great way to find out if you are looking at a "cut-and-paste program." Find any blank form in the safety program manual and ask for copies of the completed form. Many times, this exercise has resulted in the honest answer that the manual was prepared by a safety consultant, and the company is not aware of what expectations have been established by the program documents.

This moment of truth is also when we can lose understanding of the safety context of the company and current safety challenges. A checklist mentality and the discovery of a dusty cut-and-paste program manual might help us understand the true value placed on safety. The mere presence of policies, programs, and procedures does not tell us much, but the usage of the manual will tell us a great deal about safety culture. In a world where artificial intelligence can write a safety manual, how can we gauge safety culture in practice? It has to be an interactive process, much more than a questionnaire can provide. Something like the conversation I had with the office admin.

But given the short amount of time when an interaction with a contractor occurs, what are the most important questions? What if you were in an elevator and needed to learn something in the one minute you had together? What one question would you ask a contractor?

It matters when the opportunity comes to ask the question, too. For example, we want to ask different safety questions when we are in the field on an active job. But if we are in an office setting or somewhere where we can relax for a few minutes, the questions can change.

In many years, I have not seen this exact question on many checklists (it has always been on the checklist of all the organizations that I have been a part of, however). There are lots of topics that align with this question. Perhaps it's because most contractor safety questions are focused on "yes/no." For this question, you must provide a narrative response, and it's not easily graded. It's not multiple choice, or true/false. It requires a story. The best question I have ever asked a service company is:

> What are you doing to improve your safety performance?

If you pay attention to how this question is answered, it will tell you everything about a company's safety culture. Let's break down what you can learn.

> Who can and actually does answer the question?

You will learn about the company's safety culture by who can fully answer this question. Is it a safety professional or an operations person who leads the conversation? Is the answer equally understood and explained? We are looking for safety ownership. If a company looks to the safety professional exclusively for all the answers, there may be an opportunity for operations to embrace safety in a way that allows them to walk the talk. In companies with higher safety maturity, the answer will come from operations and be well understood, especially in terms of critical success factors, key performance indications, leading indicators, and focus programs.

8.2 DO YOU GET AN ANSWER WITH SUBSTANCE?

If the response is focused only on actions required by law, it could indicate a very elementary understanding of safety. If the response is focused on proactive programs, leading indicators, culture, and engagement, it will indicate a higher safety maturity. If you have to explain the question or give examples, this will generally indicate it's a new question, and that will be another indicator of safety maturity.

8.3 HOW LONG IS THE PAUSE BEFORE YOU RECEIVE AN ANSWER?

If the answer can be given readily, it will indicate how fresh and active the efforts are. Ideally, the answers could be a part of the regular morning discussions, presentations, and safety meetings. If they are frequently discussed, the information will be easily provided. If the answers come like you are pulling teeth, it's a general indicator that it's not an active initiative.

8.4 CAN EVERYONE GIVE A CONSISTENT ANSWER?

Ideally, all employees could answer this question to some degree, and you might ask this same question to different roles within the company. If the answer varies significantly, it can be an indicator of safety communication effectiveness.

8.5 HOW EASILY IS IT ANSWERED? IS IT NATURAL? DO THEY STRUGGLE TO DESCRIBE CULTURE?

If the answer comes in a fluid, coherent, and succinct form, it will indicate how safety systems are ingrained in the culture. If the efforts cannot be easily described, it will indicate an opportunity for safety culture acceptance and growth.

8.6 HOW SPECIFIC AND DETAILED ARE THE ANSWERS?

Do you hear generalities, or do you hear about programs and initiatives with substance and meaning? The more details are provided, the more confident you can be that the efforts are real and are working as intended. The more you hear about roles and responsibilities for getting the results, the better.

8.7 ARE THE ANSWERS FOCUSED ON RULES, STATISTICS, LEADING INDICATORS, OR INCIDENT LEARNING?

Ideally, improvement will be focused on leading indicators like near-miss communication.

8.8 ARE THEY EXCITED AND ENGAGED TO TELL THEIR STORY?

Gauge the energy level of the answer. How passionate they are in telling their story will tell you how engaged they are in safety improvement.

8.9 DO THEY ASK FOR YOUR ADVICE?

If they are truly interested in improving, there is a good chance that they will ask for your help. The power of collaboration is a strong way to improve the outcome.

You won't learn everything about a company's safety efforts from one question, but the hope is that this one question will be the starter question that leads to a meaningful conversation about the process of safety improvement. You will learn if they are improving in a meaningful way. I believe that there are a few better ways to evaluate the safety performance of a prospective service company.

In life, everything is a test. A test to see how we will respond. We have thousands of decisions that we make each day. Doing nothing different is a decision. The one question about how a company is improving is a test to see how people respond. Will they rise up or lay down? Will they see the test for what it is or completely miss it? It's up to them to make something of it. So, ask, ask again, ask more, and don't stop.

9 Getting Better When Nothing Happens
Proactive Learning for Clients and Contractors

Ron Gantt

9.1 INTRODUCTION

One day, while walking through a break room at a client's site, I noticed a sign on the wall. It was a simple sign that had just four lines:

CAUSES OF ACCIDENTS:

1. I didn't think
2. I didn't see
3. I didn't know

I'm someone who likes to overthink things, and this seemed ripe for overthinking. Reflect on this for a minute. How would someone determine that those are the causes of accidents?

Simple. They either experienced an accident themselves or saw someone else have an accident, and they looked back to figure out how it happened. In doing so, they saw that they (or the other person) didn't think, see, and/or know. Effectively, they did an accident investigation. Or, more likely, they did multiple investigations, and they noticed some patterns that led them to believe that all causes of accidents fit into one of those three categories.

Set aside for a minute whether you agree or disagree with the conclusions that whoever made the sign came to from their accident investigations. They did something very important that we all do during our investigations. In fact, it's so common we rarely, if ever, even think that we are doing it. They used counterfactual reasoning.

Counterfactual reasoning is pretty much what it says – it is "counter-fact." It is describing something that did not happen. In investigations, when we use counterfactual reasoning, we say that if only something else had happened or had not happened (the counterfactual), then the accident would not have occurred. Using the earlier sign, basically, the person who made the sign looked at one or more accidents, found instances of people not thinking, seeing, or knowing, and then they figured out that

DOI: 10.1201/9781003679448-9

if only those people had thought, seen, or known, then the accident would not have occurred. Therefore, the cause(s) of the accident were not thinking, seeing, or knowing.

Now, you may be thinking to yourself, what does this have to do with the topic of this chapter? That's a fair question, but here's the interesting thing about that sign – what's the intention of having a sign like that? Is it just an educational sign telling people about history? Like, they just wanted people to know what the causes of accidents were on their way back from a break?

No. That sign is a safety sign. The intention of that sign is to help people be safe. To illustrate, if we adjust the sign a bit but keep the same underlying message, that sign just as easily could say:

CAUSES OF NOT HAVING AN ACCIDENT:

1. Thinking
2. Seeing
3. Knowing

The idea is to tell people what the sign maker thinks is necessary to help people not have accidents. It's a sign about safety, not really about accidents.

And this is where we come to the point of this chapter. Whether we agree with the conclusions of the sign or not, the methods the sign uses are standard for managing safety. We know what to do for safety because we look at accidents that have occurred, try to figure out what caused them, and then use that to figure out how to stop accidents from happening in the future. You see an example of this whenever you hear someone say that safety rules or regulations are "written in blood." Someone was seriously injured, we identified a rule or regulation that we think would prevent future injuries, and then we create systems and processes to make sure those rules or regulations are followed.

It's the basic formula of safety management. Accidents lead to learning, which leads to safety practices. As a result, we in safety are obsessed with things going wrong. You could almost say that in safety we aren't really that interested in "safety," but rather "unsafety." It's far easier, for example, to identify something "unsafe" than to identify something "safe." This comes because most of our learning in safety comes from investigating accidents, learning about what is "unsafe," that is, reactive learning.

Does it have to be this way? The short answer is no. There's nothing stopping us from learning proactively. In fact, in my opinion, not engaging in proactive learning may be contributing to a lot of misleading beliefs in safety and contractor management and may be holding us back from improving our safety performance. In the rest of this chapter, I will start by describing some of the unintended consequences of learning only reactively. Next, we will explore why proactive learning is so essential to proper site management. Finally, I'll describe some of the tools and methods I've used at job sites with contractors to help learn proactively.

9.2 A VIEW ASKEW

If you had to say right now what you think the most common cause of accidents is, what would you say? If I had to guess, I'd say you mentioned something involving people. Whether it's people violating rules, not thinking through jobs, managers not devoting enough time to safety, contractors trying to cut corners, etc. This sort of thinking is common, especially in contractor management. Basically, you can boil this down to three logical assumptions:

- Our systems and processes are basically safe enough if they are used as designed.
- People make them unsafe when they deviate from them.
- Therefore, we must develop systems and processes to control them.

Now you may or may not have thought about it that way, but I imagine this looks pretty familiar. The idea is that we have things we ask contractors to do, like training, job hazard analysis, inspections, permits, etc., that, although not perfect, are good enough that if people just followed them then we wouldn't have accidents. So why do we still have accidents? People aren't doing what they are supposed to do!

Therefore, we need to get people to do what we need them to do. That can involve some harsh stuff, like discipline, zero tolerance, and what not. By contrast, sometimes we couch it in friendly terms by saying things like we are trying to build a "safety culture" or going after the "hearts and minds" of workers. But really, at the end of the day, all we are trying to do is to get people to be happy about following the rules.

This approach doesn't come from nowhere. You may have seen statistics that say things like 88% of accidents are caused by unsafe acts or 90% of accidents are caused by at-risk behavior. If that's true, then we likely get the most bang for our buck by trying to eliminate unsafe or at-risk behavior (or whatever we call it). If we can get rid of the unsafe behaviors, then the only things we are left with (almost by definition) are the safe behaviors.

The pie chart in Fig. 9.1 illustrates this a little bit more starkly. It shows a system with a failure rate of one in 10,000. This means for every instance of doing that task,

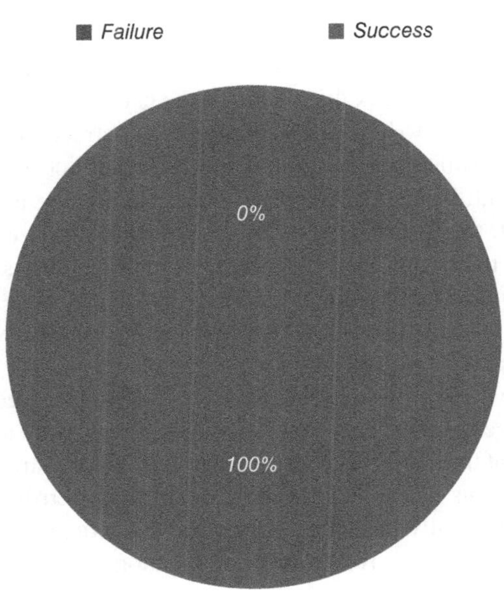

FIGURE 9.1 Failure Rate.

there's one failure, which is the small red line in the middle, with the rest being the successes. To put this in perspective, on a large data center construction project on which I was involved, we calculated the total work hours of everyone at the site, all subs and trades, and the total number of incidents (first aids, near misses, everything), and we determined that we had a failure rate of about one in over 200,000. So, one in 10,000 is a pretty high failure rate.

The reactive learning-only approach to safety is saying that all we have to do is figure out what's happening in that one, tiny, thin sliver of failure in the center. Of course, focusing on this one data point means ignoring what's happening the rest of the time, and if we do that, all we will be left with is the rest of the circle without any bad consequences. Does that make sense to you?

It shouldn't. The idea that all we need to do to understand safety is to look at accidents is basically the same as saying all we need to do to understand marriage is to look at divorce. Sure, if you want to understand marriage, you should look at divorces. But that is only a small part of the overall experience of marriage that people have. Only looking at failure doesn't mean you understand success. Avoiding divorce doesn't mean you have a happy marriage. Only looking for unsafe things doesn't mean you know what it takes to create safety day in and day out at your job site.

And this leads to a much scarier point – only looking at failure doesn't mean you understand failure. Think back to the sign we talked about earlier – the causes of an accident are not thinking, seeing, or knowing. Think about where you are right now reading this. Are you thinking, seeing, or knowing about everything that can hurt you in the room you are in right now? Of course not! In fact, if you think about it, you probably thought of hardly any of the things that could hurt you. The chair you're sitting in. The building you're in. The coffee you are drinking. All of it could hurt you in some way, and you likely only did cursory things to protect yourself from those harms. And yet you aren't having an accident right now and are highly unlikely to have one.

Now, this isn't to say that not thinking, seeing, or knowing are good things. It's just to say that it is easy to think that these are the causes of accidents if all we do is look at accidents. It's easy to look at an instance of a contractor bumping into another piece of equipment while backing up and conclude that not paying attention is the cause of the accident, because if only they had been paying attention to the other piece of equipment, they wouldn't have hit it. But this doesn't explain how the person was managing their attention in the midst of multiple things competing for that attention in a dynamic worksite. Not understanding what is happening when nothing is happening means we misunderstand what happens when the worst things happen and therefore end up trying to solve the wrong problems.

In fact, I believe the reason we think most accidents are caused by "human error" or "unsafe acts" is because the only time we look is when we are looking at accidents. We don't realize that the same behaviors that we call "errors" are, in another context, successful actions. This means that those statistics about "errors" and "unsafe acts" we saw earlier are basically meaningless because the actions themselves are not the problem, but the actions in a given context or situation. Let's explore this idea some more by asking a critical question.

9.3 WHAT'S HAPPENING WHEN "NOTHING" IS HAPPENING?

To understand this further, we should take a step back and ask what's going on normally, in those 9,999 times when things go well. Think about your job sites where you work with contractors (or if you are a contractor, think about where you work with your client). Do any issues on the list below sound familiar?

- Outdated, inapplicable, or missing procedures
- Inadequate training
- Aging tools and equipment
- "Temporary" fixes or solutions that have been in place basically permanently
- Poor communication, particularly between different organizations
- Unrealistic plans and schedules
- Insufficient resources (e.g., money, time, people, etc.)
- Missing or out-of-touch leadership

If you're working at most job sites I've worked at, at least half of the abovementioned list resonates with you, if not the whole list. The next perfect job site I go to will be the first one, and I'm sure your situation is the same. It makes me wonder how any of us get anything done in such conditions. But we do. Day in and day out, we find a way to get work done even in the face of such imperfections.

Here's the thing that should scare us – if you look at publicly available accident reports for major accidents, you know what you find consistently listed as causal factors? The things on the abovementioned list. If we look at major construction accidents, aviation crashes, refinery fires, and explosions, stuff you hear about in the news, you will find that investigators point to the things on the list that we all know are present at our job site *today* as factors that caused or contributed to the catastrophic event.

A lot of times, as a safety professional, I get asked the question, "why are we having these accidents?" It's a valid question most of the time, and definitely a question we should work to have an answer to. But there's another question that almost never gets asked, but when you think about the abovementioned list and how it relates to catastrophic events, maybe we should ask it.

9.3.1 Why Aren't We Having More Accidents?

Think about it. Why aren't we killing more people at our job sites? If you had to answer this question in a clear, definitive way, could you? My concern is that we don't really have a good answer to this question. But the answer to this question is critical because the answer to this question is . . . safety.

Now some of you are thinking that the answer to the question is luck (or if you don't like the term "luck" you can use the fancier term "stochastic statistical variance," which basically means the same thing but makes you sound smart and fancy). But luck can't be the only factor making it so when we perform a task, we can feel confident 99.99% of the time we won't have an accident. If you're that lucky, you should quit your job and gamble for a living. It can't be only luck.

Is it because of our safety management systems or processes? Maybe. But we already admitted that those are flawed. So, it can't only be that either.

9.4 WHAT ELSE IS THERE THEN?

We already identified it earlier when we talked about how you get work done despite the imperfections at your job site. It's people. Specifically, the ability of people to adapt to imperfect conditions and reliably achieve success. People find a way to pull success from the jaws of failure day in and day out, and they do it so skillfully and reliably that we don't even realize it's happening. In fact, people are so good at adapting to the imperfections of our work systems that when it fails, we often blame them for it.

Take, for example, a contractor coming to your job site and attempting to start work for the day. Simple task, right? But that contractor must find a way to figure out what's required of them. Often, they need to talk to their supervisor to figure out what work they will do that day, which should be relatively straightforward. But each job site almost always has its own set of requirements related to pre-task planning and paperwork. So, that needs to be done. Also, every job site is a little bit different, so they can't just rely on their memory to help them do it right. There are nuanced differences they must pay attention to that could be easy to miss. Sure, the basic points are the same, For example, identify hazards and mitigations, but sometimes they forget a step in the paperwork or the process itself because they didn't have to do it that way at their last job or it's not something they are used to.

Of course, we go over how to do all of this in the job site orientation, so that should be sufficient to get them up to speed so they can perform reliably. If we're honest though, our orientations aren't that great. It's a lot of information thrown at people all at once, most of it isn't relevant to their task, and it's often delivered in a very boring way.

This is a really simple example, and we haven't even fully unpacked the challenge this pre-task task holds for the contractor, but notice how there are a lot of opportunities for failure here. And reflect on how little of the time things go wrong. Usually, the pre-task paperwork and planning is done mostly correct. When it isn't correct though, we point the finger at the contractor and tell them to do better next time. We might even send them through the same poor orientation that we put them through the first time, as if expecting different results from the same actions is not part of the definition for insanity

We obviously can do much better. But doing better starts with recognizing first and foremost that the three assumptions mentioned earlier need to be updated a bit. The updated version of them is:

1. Our systems and processes are always flawed, and we aren't sure exactly how flawed they are.
2. People fill the gap between our flawed systems and processes and reality in order to get work done successfully.
3. Therefore, we should develop systems and processes to learn about those gaps and figure out ways to support people better.

These three assumptions form the cornerstone of the proactive learning perspective. To get better at contractor management and to be able to answer the question "why we aren't having more accidents," we need to get better at learning when "nothing" is happening. The rest of this chapter will describe some ways to get started doing that, but step one is recognizing that it's necessary. We don't have to wait for accidents to happen to figure out how to improve things. Even our audits and inspections often miss important factors about the work happening because they often are based on comparing the work to what *should* happen. That's important, but it's not complete. We not only need to compare the work we see to what *should* happen, we also need to understand what *is* happening. Our audits and inspections may help us with this but only if we come at them from a different mindset. So, the first step in starting to learn proactively is to adopt that mindset and that's what we'll talk about next.

9.5 TREAT EVERYTHING AS A SYMPTOM

To get started in proactive learning, make this mindset shift – move from a mindset that evaluates work for compliance with some rule or process to one that is curious. This isn't to say that compliance is bad. There are times for compliance inspections and audits. But proactive learning is different.

One of the fundamental differences between audits and proactive learning comes from what we described earlier – the adaptability of people. But we must remember that people don't adapt randomly. If someone is adjusting to what they are doing, they are adjusting to something. Often that something is not as easy to see as the behavior we are observing (i.e., the adaptation). This is why we typically focus on the behavior or the person to fix the problem. But a proactive learning mindset starts by admitting that what we are seeing is a reaction to something else. One way I put it is that *the behavior is a symptom* of something.

To illustrate, think of the pre-task planning example we used earlier. If we see a contractor not filling out their pre-task plan paperwork correctly, that is definitely a problem. But the proactive learning mindset does not stop there. It sees the behavior (i.e., not filling out the paperwork correctly) as a symptom of something else. Is it confusion over how to fill out the paperwork? Is it a lack of understanding of the purpose of the paperwork? Is it lack of time to fill out the paperwork? Maybe it's something else entirely. So, the proactive learning mindset begins with being curious about the thing we are seeing and uses it as the start of the learning process, rather than the end of it. The idea behind this is that in a proactive learning mindset, we want to make sure we are solving the right problems, not just fixing the symptoms of the problems.

One practical tool you can use to start creating this mindset in yourself or others is to change the way you ask questions about the problem you are trying to solve. For example, if you think of what we described earlier about how messy and imperfect work is, you might shift your question to focus in on the challenges people face when doing this task. In the pre-task planning paperwork example, we often start by asking the question, "why isn't the contractor filling out the paperwork correctly?" As an alternative, though, you might shift the question away from the contractor toward the systemic issues and challenges that might be influencing them. "What makes filling out the paperwork correctly challenging?" Asking the question this way turns the focus toward your system and areas where people must overcome the imperfections in your systems and processes to be effective. This doesn't mean you have to fix all those things, but you might find areas where you can add more support or make it easier on people, which makes it easier to get the outcome you wanted.

9.6 PROACTIVE LEARNING TOOLS

Once you have a proactive learning mindset, based on curiosity, you also likely will want to utilize some of the proactive learning tools to help you take the next step. In this chapter, we won't be able to exhaustively give you a "how to" for every proactive learning tool because there just isn't the space for that, and you might also find that you can use what we've talked about to come up with your own tools or adapt some other processes in your organization to help you get the information you want for proactive learning. I should also note there are plenty of other resources out there that go into more detail on certain tools, particularly from authors like Todd Conklin, Bob Edwards, and Andrea Baker, who talk about things like learning teams or operational learning. I would highly recommend anyone who wants to dive deeper into proactive learning also utilize the works of these and other authors to help them get started. For the rest of this chapter, though, I want to talk about two other tools I've used with contractors at job sites I've worked on that are easy to describe and implement and useful for proactive learning.

Before we get too far, I also want to note that a critical part of any "learning" is changing things. That means any proactive learning process needs to include follow-up on any actions that come out of the process. If proactive learning isn't, at least most of the time, leading to actions that improve the job site, then you may want to evaluate what you're doing. In my experience, a lot of improvement is often a function of one of the following:

1. Not utilizing the learning tool in the most effective way (i.e., making it more like an audit or not asking the right questions).
2. Not capturing actions from the learning tool in the site's action tracking system.
3. Not following up on actions to discuss whether they have been implemented and have improved things.

One thing I've found useful is to have proactive learning as one of the topics discussed in the regular meetings we have with our contractors, so we can discuss the outcomes of the latest proactive learning events, actions that came out of them, and the results we are seeing. Just making it a regular point of discussion sort of builds in visibility and accountability into the process and often gets things done, in my experience.

9.7 LEADER STOP & CHATS

The first proactive learning tool I've found success with that is really easy to implement is something I call "Leader Stop & Chats." You may have a process where your site leaders go out and do site walks already, and you may call them different things, like Leader Site Walks, Gemba Walks, etc. The Stop & Chats are similar to that, but one difference is that the purpose of these Stop & Chats is not to coach workers for compliance or set an example for safety.

Again, those are worthy goals, but they just aren't the goal of the Stop & Chats. The goal of the Leader Stop & Chats is to get leaders out in the field, directly engaging with craft workers about the challenges they face and learning about the messy details of work for the purpose of not only fixing issues in the field, but also having leaders learn about the work itself. So, the key intervention is actually in the leader (i.e., they are learning something), rather than the worker.

You may already be doing this or have a version of this at your job sites. If so, that's great. Keep doing it. However, in my experience, most people do not do this. Or if they do it, it is just leaders walking around and looking for obvious safety or compliance issues and fixing them in the field. Remember, though, the proactive learning mindset starts with seeing everything as a symptom. So, if your leaders aren't using their observations as the beginning of the learning process, then they probably aren't really using their site walks for proactive learning. It's probably closer to an audit or inspection.

One thing I emphasize to highlight the difference is in the title – Stop & Chat. The idea is that leaders go up and talk to workers. If they aren't engaging with the worker, then they probably aren't getting what they need from the site walk.

In the case of proactive learning, Stop & Chat also has a different tone. It starts with trying to understand the work and how challenging it is. That's why when we train our leaders to do this, we often recommend they start the Stop & Chat by introducing themselves and then just asking "what are you working on today?" Doing that not only tends to calm the worker down (they often feel more comfortable talking about their task than answering questions about safety), but it also focuses the conversation on the task itself. As the worker describes what they are doing, we train our leaders to ask follow-up questions specifically about where the work might be difficult (e.g., "that piece looks like it's wedged in there tight. How do you get it out?); what pressures they may have (e.g., "how long does a task like this take? How much time were you given to do it?"); what resources they might need (e.g., "what tools do you need to do this task and do you ever have any trouble getting them?); and what other challenges they may face (e.g., "I see you're doing this work right near where they are pressure testing those pipes. How do you coordinate your work with theirs so both crews can do what they need to do?").

We recommend keeping the conversations short, usually not more than 5 minutes, so the worker can get back to work. And often, in that time, you find that you get enough information for the leader to chew on. Sometimes you'll find that you'll have action items right away that you can follow up on (e.g., the workers mention they need a specific tool that would make their job easier and safer). Often, you'll have some things you may want to follow up on to figure out where the problem is coming from (e.g., the workers report problems with other crews working in their area in a way that makes it more challenging for them, but both crews are working where they were assigned). In some cases, you just learn some information that could help you understand how decisions made away from the field are affecting the field. For example, one leader I went on a walk with talked with a worker who told them their work was easy unless someone decided to reduce the time they had to get it done. As we were walking away, the leader admitted that before they talked with those workers, they were just about to reduce the amount of time in the schedule for them. Now the leader had more information about the consequences of that decision, and they could make a more informed decision.

In any case, after walking away from the interaction, we train our leaders to go and capture what they learned, particularly any actions they had. At one site I worked at, at the daily planning meeting, leaders who did site walks would report what they found and any actions they wanted to follow up on. You may find you want to record it differently at your job site. Use what works for you.

One last note on this tool. Note that, in the previous section, I have mentioned several times how we trained our leaders. Yes, we are talking about just going and talking with workers, so it's easy to assume that leaders already know how to do that. Don't make that assumption. In my experience, having a short training session and then going out with them once or twice to coach them is vital. It also might help to give them a cheat sheet. However, I strongly advise against giving them a set checklist or similar to use. The last thing you want is them reading a script or a checklist while talking to a worker. This should be a human conversation, not a check-the-box exercise.

9.8 RAPID INSIGHTS

One other tool that is easy to deploy for proactive learning that I'd recommend is something we have taken to call Rapid Insights. The idea behind Rapid Insights is that it is a quick way to get some information about how things are really working in a specific area you may be interested in. You can use it for any topic of interest. At our job sites, we primarily use it to learn about our high-risk activities, but we've also used it for things like how our construction and contractor teams are coordinating.

9.8.1 Here's How You Do It

The first step is to get a small group of people (no more than 12, no less than two) who are doing the task you're interested in learning about. So, for example, if you're interested in learning about how electrical work is going at your job site, you'd bring in a group of electricians. You can do this in a conference or training room or even out in the field if you want to.

Second, after introducing yourself and explaining what you're doing, you can ask the group to identify the small number of critical items needed for the task to go well. In the case of high-risk activities, we ask, "when you're doing this task, what are the control measures that stop you from dying or being critically injured?" When we inquired about coordination, we asked "what are the critical things that need to be in place for us all to work well together?"

The key here is to focus on critical things. Yes, there are many important things, but the number of critical things is very small (two to five usually). For example, safety glasses are important for a lot of tasks, but they rarely stop people from dying and therefore aren't critical. So, when people bring up noncritical items, it's ok to push back on them a bit. Don't worry too much about if they don't bring up all the things you think are critical (unless they don't mention something that you really want to know about), or if they mention something that you aren't sure is really critical. If they push back on you, that they believe it's critical, just note it and move on.

If you're able to, it helps to write each critical item on something so everyone can see, like a dry-erase board. Then, the third step is to have each person in the group rate each critical item on a scale from 0 to 4 (0 is terrible, 4 is excellent) based on how that item is going at your job site. So, for example, let's say the electricians say that lockout/tagout is a critical item. We then would ask them to rate how the lockout/tagout processes at our job site are. It's important to have each person rate it individually. If you're worried about power dynamics in the room, you might also consider having each person write down their answer on something, such as sticky notes, before asking for everyone's answer. If you're using something like a dry-erase board, jot down everyone's answer on the board. Do the same thing for each of the critical items. At the end of this step, you have a board with critical items listed and the ratings each person gave for each of those critical items.

The fourth and final step is to talk about the ratings. Note that you likely don't have time to talk about each one. So, focus on those areas where either everyone rated something low or there is disagreement (i.e., some rated it high and others lower). Go through each and ask, "what's going on here?" Often, they will tell you stories or give you examples where the critical item isn't working as it should. For example, we've heard stories where employees who were supposed to use fall protection in a given area identified specific places where it was impossible for them to comply with the requirement. This allowed us to ask follow-up questions about where those areas are and what about them is making it difficult or impossible.

That's it. You can let everyone go back to work at that point. The process takes around 15 minutes. But at the end, you have a set of stories where things are not working as well as you'd like them to, things that you can then go and explore or take action, as appropriate.

One common mistake people make with rapid insights is trying to track the ratings or numbers. It's important to remember that the numbers are essentially meaningless. We are only using them as a conversation starter to identify differences from expectation. For example, you may have one critical item where everyone rates it as 4 except one person who rates it a 3.5. Looks good, right? Well, that 0.5 difference can indicate a serious problem that you want to dive into. Remember, what is a 3 or 4 to one person may be something else to another person. The outcome of the rapid insight is not the rating. It's the examples and stories where things aren't working as well as we'd like that you can go and work on.

9.9 CONCLUSION

If you are reading this book, then you already know that working with contractors is hard, and you probably don't need me to tell you that. It's messy. It rarely goes as planned. And that's why we can't just assume that we can put forth a set of expectations and expect everyone to follow it without problems. A proactive learning approach helps us get feedback on how our set of expectations is working in practice. It helps us learn what works and what doesn't, so we can get better and better at making our job sites safe and effective. Yes, it takes time and a bit of effort. But the rewards we get are easy to see in not only better safety, but also less wasted time doing stuff that just isn't working, better engagement, and just, in general, a better job site.

10 Contractor Hazard Identification and Risk Assessments

Josh Ortega

10.1 INTRODUCTION

Keeping contractors safe on the job is important for any company. Contractors often work in different environments and need to understand the risks involved. Identifying hazards and assessing risks properly can prevent accidents, protect workers, and ensure legal compliance. This chapter explains how to find and manage hazards to keep workplaces safe and productive. Additionally, it discusses the tools and methods available to enhance safety protocols, including leadership engagement, human and organizational factors, and the role of technology.

Contractor safety is a shared responsibility between the contracting company and the organization hiring them. Effective hazard identification and risk assessment (HIRA) ensures that both parties understand potential dangers and implement controls to prevent injuries, fatalities, and financial losses. Without proper hazard assessments, companies risk facing legal consequences, reputational damage, and project delays due to workplace incidents.

10.2 WHY IDENTIFYING HAZARDS AND ASSESSING RISK IS IMPORTANT

Contractors may not be familiar with the job site, making them more likely to experience accidents. They may also use different tools and techniques that introduce new risks. Many workplace accidents occur due to human mistakes and systemic problems (Dekker, 2011). Using a structured HIRA approach, companies can prevent accidents before they happen.

A strong hazard identification process ensures that employees and contractors are aware of the potential dangers and that proper mitigation strategies are in place. Effective risk management improves productivity, decreases accident-related costs, and helps maintain compliance with safety regulations.

Poor hazard identification and risk assessment can lead to major workplace disasters. The 2010 Deepwater Horizon explosion is a tragic example, where failure to identify hazards contributed to an explosion that killed 11 workers and caused a massive environmental disaster (National Commission on the BP Deepwater Horizon Oil

DOI: 10.1201/9781003679448-10

Spill and Offshore Drilling, 2011). Proper hazard identification could have mitigated risks and prevented this catastrophic event.

Beyond catastrophic events, poor hazard assessments also contribute to more frequent but less severe incidents, which can increase costs, disrupt operations, and lower morale. Companies must proactively assess risks at every stage of contractor engagement, from prequalification to project completion.

10.3 KEY CONTRACTOR MANAGEMENT CHALLENGES IN RISK ASSESSMENT

Contractor risk assessments involve unique challenges that differ from managing employee safety. These include:

- **Language and Communication Barriers:** Contractors may have limited proficiency in the primary language used at the worksite, affecting their understanding of safety protocols. Miscommunications can lead to errors in hazard identification, improper use of personal protective equipment (PPE), and difficulty in following emergency procedures. Organizations should consider multilingual training programs and visual safety aids to bridge these gaps.
- **Varied Safety Cultures:** Contractors often work with multiple organizations, each with different safety expectations and training levels. A lack of alignment between the contractor's safety culture and that of the hiring organization can lead to inconsistencies in hazard recognition and compliance with safety measures. Conducting a thorough cultural safety assessment and reinforcing shared safety values can help mitigate this issue.
- **Subcontractor Oversight:** Ensuring that subcontractors follow safety standards is difficult when working within complex supply chains. Subcontractors may not be directly vetted by the hiring company, leading to gaps in safety compliance. Companies should implement strict onboarding processes and require subcontractors to meet the same safety qualifications as primary contractors.
- **Short-Term Work Assignments:** Many contractors work on temporary projects, limiting their familiarity with site-specific hazards and risk controls. This lack of familiarity increases the likelihood of incidents occurring due to improper hazard identification or failure to adhere to established safety protocols. Providing concise but comprehensive safety briefings before each shift can improve awareness.
- **Regulatory Differences:** Contractors operating across different countries or regions must adhere to varied safety regulations, making compliance more challenging. Organizations should develop a centralized compliance framework that accounts for international regulatory requirements and ensures consistency across all worksites.
- **Emerging Risks:** Advancements in technology and automation introduce new risks that require continuous safety updates and risk assessments. For

example, increased reliance on robotics and AI-driven machinery presents potential hazards related to system malfunctions or human–machine interactions. Conducting regular safety audits to assess the impact of technological changes is essential for maintaining a proactive risk management approach.

- **Remote and Isolated Work Locations:** Some contractors work in remote or high-risk environments, where access to emergency assistance and resources is limited. Harsh weather conditions, long work hours, and psychological stressors can further elevate risks. Employers should establish strict emergency preparedness plans and provide remote communication tools to ensure worker safety in isolated job sites.
- **Human Factors and Worker Fatigue:** Contractors often work under high-pressure conditions with demanding schedules, increasing the risk of fatigue-related accidents. Fatigue impairs cognitive abilities, reaction times, and overall hazard awareness. Implementing mandatory rest breaks, fatigue management training, and scheduling flexibility can help reduce fatigue-related incidents.
- **Security and Unauthorized Access Risks:** Large projects or highly regulated industries may face security risks related to unauthorized personnel entering restricted areas. Poorly managed access control systems can expose workers to safety threats. Companies should use advanced security protocols, such as biometric access control, ID verification, and real-time monitoring systems, to mitigate unauthorized access risks.

To address these challenges, organizations should implement a structured, risk-tiered contractor safety assessment process, regularly review safety performance data, and engage contractors in proactive safety improvement initiatives.

10.4 RISK-BASED CONTRACTOR SELECTION APPROACH

Not all contractors carry the same level of risk. Companies should categorize contractors based on the complexity and exposure of their work, using a risk-tiering system:

- **Low-Risk Contractors**: Office-based roles, administrative work, and consultants.
- **Medium-Risk Contractors:** Equipment maintenance, routine fieldwork, and nonhazardous operations.
- **High-Risk Contractors**: Construction, electrical work, confined space entry, and hazardous material handling.

Each risk tier should have distinct levels of prequalification, safety training, and monitoring requirements. High-risk contractors should undergo more rigorous safety evaluations, while low-risk contractors may require a simplified process. Competency verification strategies, such as hands-on testing and behavioral assessments, should be integrated into the selection process.

10.5 HUMAN AND ORGANIZATIONAL FACTORS IN HAZARD IDENTIFICATION

Understanding human and organizational factors is critical to improving hazard identification and risk management. These factors influence worker behavior, decision-making, and the effectiveness of safety programs. Key considerations include:

- **Safety Culture:** Organizations must foster a culture where safety is a priority, and workers feel empowered to report hazards without fear of retaliation.
- **Leadership Engagement:** Strong leadership commitment to safety ensures that hazard identification remains a priority at all levels of the organization.
- **Training and Competency Development:** Continuous safety training equips contractors with the knowledge and skills needed to recognize and mitigate hazards.
- **Behavior-Based Safety (BBS)**: Encouraging safe behaviors through observation, feedback, and positive reinforcement helps reduce human errors and improve safety outcomes.
- **Incident Reporting and Learning Systems:** Effective feedback mechanisms enable organizations to analyze incidents, identify trends, and implement preventive measures.
- **Cognitive Biases in Risk Perception**: Understanding how workers perceive risks can help improve training and hazard communication.

10.6 ADDRESSING PSYCHOSOCIAL HAZARDS IN CONTRACTOR MANAGEMENT

Psychosocial hazards are work-related factors that may cause psychological harm, and they can have a significant impact on contractor performance, engagement, and safety outcomes. ISO 45003:2021 defines psychosocial hazards as aspects of work organization, social factors, and work environment that have the potential to cause psychological or physical harm (ISO, 2021).

10.7 COMMON AND MANAGEABLE PSYCHOSOCIAL HAZARDS

For organizations managing contractors, several frequent psychosocial risks are relatively easy to identify and control:

- **Lack of role clarity:** Contractors often report confusion about their responsibilities or who to report to. This can be mitigated with clear, written scopes of work and structured onboarding briefings (ISO, 2021; Leka & Jain, 2010).
- **Poor communication and feedback:** Infrequent or unclear communication can lead to frustration, mistakes, and reduced morale. Simple actions, such as regular check-ins, daily toolbox talks, and encouraging open dialogue, can improve clarity and connection (Safe Work Australia, 2019).

- **High workload or time pressure:** Unrealistic schedules or frequent urgent demands contribute to stress and burnout. Organizations can help by involving contractors in scheduling decisions and building reasonable lead times into scopes of work (Cox et al., 2000).

These hazards are relatively simple to manage, yet they often go unaddressed, especially in fast-paced or contract-heavy environments.

10.8 MANAGING PSYCHOSOCIAL RISKS: PRACTICAL INTEGRATION

Organizations should incorporate psychosocial risk controls into existing contractor safety practices. Key actions include:

- Including psychosocial risks in pre-task risk assessments or job hazard analyses (ISO, 2021).
- Providing mental health awareness training to supervisors and site leads.
- Offering equitable access to support services like employee assistance programs (EAPs), even for short-term contractors.
- Reducing isolation risks by checking in on remote or lone workers, especially in remote, Fly-in, Fly-out (FIFO), or offshore arrangements.

Worker participation is essential. Contractors should be encouraged to report psychosocial concerns without fear of retaliation, and their input should be integrated into continuous improvement efforts.

10.9 ORGANIZATIONAL BENEFITS

Actively managing psychosocial hazards leads to better engagement, fewer safety incidents, and improved contractor retention – outcomes that directly support project success and compliance with ISO 45001 and ISO 45003 (ISO, 2021).

10.10 RULES, STANDARDS, AND TOOLS FOR CONTRACTOR SAFETY

Ensuring contractor safety is critical for businesses that rely on third-party workers. Contractors often operate in environments where they may not be familiar with company policies, site hazards, or regulatory requirements. Implementing a structured approach to contractor safety management helps mitigate risks, improve compliance, and promote a culture of safety. Several organizations provide guidelines and tools to help companies maintain high safety standards.

10.11 KEY REGULATIONS AND STANDARDS FOR CONTRACTOR SAFETY

Various national and international regulations and consensus standards establish requirements and guidance for contractor safety, including:

- **Occupational Safety and Health Administration (OSHA)**: OSHA's regulations mandate that employers identify and control workplace hazards. Companies must ensure that contractors receive adequate safety training and comply with job-specific requirements. Noncompliance can result in legal consequences, financial penalties, and workplace incidents.
- **ISO 45001**: This international standard provides a structured framework for organizations to develop and maintain an effective occupational health and safety management system. ISO 45001 emphasizes risk-based thinking, worker participation, and continuous improvement in safety programs (ISO, 2018).
- **ANSI/ASSP Z10**: The American National Standards Institute (ANSI) and the American Society of Safety Professionals (ASSP) collaborated to develop ANSI/ASSP Z10, which outlines best practices for contractor risk management, hazard identification, and control implementation.
- **National Institute for Occupational Safety and Health (NIOSH)**: NIOSH offers research-backed recommendations to help organizations enhance workplace safety. It promotes hazard prevention strategies and worker engagement in identifying potential risks.

These regulatory frameworks provide guidelines for managing contractor safety, ensuring that companies create structured programs to protect workers and reduce liability risks.

In addition to regulatory standards and consensus standards, companies can leverage various tools and technologies to enhance contractor safety management. The right tools help streamline compliance, improve hazard identification, and ensure that contractors meet safety requirements before stepping onto a worksite.

- **Veriforce Contractor Management Systems**: This tool assists companies in prequalifying, monitoring, and managing contractor safety. It helps organizations ensure that contractors comply with industry-specific safety requirements, track training records, and implement ongoing performance evaluations.
- **Digital Permit-to-Work Systems**: Many organizations have transitioned from paper-based work permits to digital platforms that improve tracking, documentation, and authorization of hazardous work activities. Digital systems enhance real-time communication and reduce administrative burdens.
- **Job Hazard Analysis (JHA) and Risk Assessment Software**: Software solutions designed for conducting JHAs and risk assessments enable companies to systematically evaluate worksite hazards, assign appropriate controls, and document findings efficiently.
- **Wearable Safety Technology** Devices such as smart helmets, connected vests, and biometric monitoring tools help track worker health indicators, alert supervisors to unsafe conditions, and improve overall contractor safety.
- Artificial intelligence **(AI) and Predictive Analytics:** AI tools analyze historical safety data to identify trends and predict potential hazards before they occur (CloudSDS, 2024). Predictive analytics can help organizations proactively implement corrective measures and improve overall safety

performance. By incorporating these tools into their contractor management programs, organizations can enhance hazard identification, improve communication, and ensure greater compliance with safety regulations.

10.12 BUILDING A STRONG CONTRACTOR SAFETY PROGRAM

Implementing a comprehensive contractor safety program involves several key steps:

- **Prequalification and Selection**: Before hiring contractors, companies should evaluate their safety records, certifications, and past performance. A rigorous vetting process helps ensure that only contractors with a strong commitment to safety are selected.
- **Training and Orientation**: Contractors must receive training on workplace-specific hazards, emergency procedures, and safety expectations. Conducting thorough onboarding sessions ensures that workers understand safety requirements before beginning their tasks.
- **Hazard Identification and Risk Assessment**: Worksite hazards should be continuously identified, assessed, and mitigated. Companies should require contractors to participate in pre-task hazard assessments and use digital tools to document risk management strategies.
- **Continuous Monitoring and Auditing**: Safety performance should be regularly reviewed through audits, inspections, and contractor evaluations. Establishing clear accountability measures helps maintain high safety standards throughout a project. Conduct post-job reviews to assess contractor safety performance and refine hazard identification strategies.
- **Incident Reporting, Response, and Learning**: A well-defined incident reporting process ensures that near-misses and workplace accidents are promptly documented and investigated. Lessons learned from incidents should be used to refine safety protocols and prevent future occurrences.
- **Safety Culture and Worker Engagement**: A strong safety culture requires active participation from leadership, supervisors, and contractors. Encouraging open communication about safety concerns and recognizing workers who demonstrate exemplary safety practices can enhance overall workplace safety. It also integrates safety expectations into contract agreements.

10.13 HOW TO IDENTIFY HAZARDS IN THE WORKPLACE

Finding hazards is the first step in making a workplace safe. Some useful methods include:

- **Job Hazard Analysis (JHA):** Looking at each step of a job to find risks and implementing preventative measures (Manuele, 2014).
- **Pre-task Risk Assessments (PTRA):** Checking for risks right before starting work to ensure safety precautions are in place.

- **Observation Programs:** Encouraging workers to report unsafe conditions and providing structured channels for feedback.
- **Using Technology:** Digital tools and AI help detect patterns and potential risks, allowing companies to respond to hazards before they cause harm (Grote, 2012).
- **Safety Walks and Inspections:** Regular walkthroughs to spot hazards that may not be evident in paperwork assessments.
- **Employee and Contractor Feedback:** Gathering insights from those on the front lines to identify emerging risks that may not yet be documented.

10.14 WAYS TO ASSESS RISK

There are several ways to assess risk, including:

- **Hierarchy of Controls:** Removing hazards, using protective equipment, and other methods to reduce risk in order of effectiveness (CDC, 2020).
- **Failure Modes and Effects Analysis (FMEA):** Finding weak points in a system and fixing them before they result in failure.
- **Bowtie Analysis:** Understanding what causes hazards and how to stop them through visualization of cause-and-effect relationships (Reason, 1997).
- **Dynamic Risk Assessment:** Adjusting safety plans as conditions change in real time, rather than relying solely on preplanned assessments (Hopkins, 2019).
- **Quantitative Risk Analysis (QRA):** Using statistical methods to evaluate potential risks and their impact on operations.

Advanced techniques like Monte Carlo simulations are increasingly used to assess risks in complex environments. These simulations generate thousands of potential scenarios to predict risk probabilities, helping organizations make informed safety decisions.

10.15 HOW AI AND TECHNOLOGY HELP WITH RISK ASSESSMENT

Modern technologies make it easier to assess risks and prevent accidents, including:

- Using past safety data to predict hazards and prevent accidents before they happen.
- AI-powered tools that analyze risks in real time and provide immediate alerts when unsafe conditions are detected (Hollnagel, 2014).
- Digital systems that track safety compliance automatically, ensuring all workers meet the required safety standards.
- Wearable safety technology that monitors environmental conditions and worker health, providing alerts for high-risk situations.
- Automated reporting systems that streamline incident documentation and reduce response times.

10.16 THE ROLE OF TECHNOLOGY IN RISK MANAGEMENT

Modern technologies significantly enhance contractor safety by improving risk assessment, compliance tracking, and real-time hazard identification. These innovations not only streamline safety protocols but also enable proactive risk mitigation strategies. Key advancements include:

- **AI-Powered Predictive Analytics**: Leveraging machine learning models and historical safety data, AI can predict potential hazards and recommend preventive actions before incidents occur (CloudSDS, 2024). These systems continuously learn from past safety trends to refine risk mitigation strategies.
- **Digital Permit-to-Work Systems**: These cloud-based platforms enable organizations to track, document, and authorize hazardous work activities in real time. Digital permits ensure contractors comply with safety protocols before starting tasks and allow instant updates when conditions change (CDC, 2020).
- **Wearable Safety Technology**: Devices such as smart helmets, biometric monitoring vests, and connected gloves provide real-time physiological and environmental data. These wearables detect fatigue, heat stress, and exposure to harmful substances, alerting workers and supervisors before conditions become critical.
- **Blockchain for Safety Compliance**: Blockchain technology enhances transparency by securely storing and verifying contractor credentials, safety records, and compliance certifications. This decentralized approach reduces fraud, streamlines audits, and ensures all workers meet safety qualifications. However, it may complicate general data protection regulation (GDPR) and privacy law compliance.
- **Digital Twins**: Virtual replicas of workplaces allow organizations to simulate work environments, identify potential hazards, and test mitigation strategies in a risk-free setting. Digital twins integrate real-time data from Internet of Things (IoT) sensors, providing an accurate picture of evolving workplace risks.
- **Automated Safety Inspections and Drones**: AI-powered drones and robotic inspection systems can perform real-time safety audits, detect structural weaknesses, and assess high-risk areas without exposing workers to danger. These tools enhance monitoring efficiency and reduce human error.
- **Augmented and Virtual Reality (AR/VR) Training**: AR and VR technologies provide immersive safety training for contractors, allowing them to practice hazard identification and emergency response in simulated high-risk environments. These training modules improve retention and preparedness.
- **IoT-Enabled Worksite Monitoring**: The IoT integrates sensors throughout worksites to continuously monitor air quality, noise levels, chemical exposure, and worker movements. Real-time alerts enable proactive intervention before hazards escalate.

These technological advancements are transforming contractor safety management by improving efficiency, reducing response times, and fostering a data-driven approach to workplace safety. Companies adopting these innovations are better equipped to prevent incidents and ensure compliance with evolving regulatory standards.

10.17 CONCLUSION

Finding and managing hazards is key to keeping contractors safe. Companies that implement structured risk assessments, leverage innovative technology, and adhere to safety regulations foster a proactive safety culture that benefits all stakeholders. Effective contractor safety programs not only reduce workplace accidents but also enhance operational efficiency and corporate reputation.

In recent years, psychosocial hazards have gained well-deserved attention as essential components of risk management. Stress, poor communication, job ambiguity, and social isolation can impact contractor well-being just as significantly as physical hazards. Organizations must therefore take a holistic approach to hazard identification, addressing both the visible and invisible risks present in contractor work environments. Integrating psychosocial risk management – as outlined in ISO 45003 – into contractor programs improves engagement, reduces errors, and supports better mental and emotional health outcomes.

A strong safety culture begins with leadership commitment and extends to every level of an organization. Senior management must set clear expectations, provide the necessary resources, and encourage open communication about hazards and risk mitigation. Regular training sessions, safety drills, and continuous learning initiatives ensure that contractors remain aware of evolving risks and best practices.

Collaboration between employers and contractors is essential for maintaining high safety standards. Establishing partnerships based on transparency, shared accountability, and mutual respect allows for more effective HIRA. Organizations should also leverage data analytics and AI-driven insights to monitor patterns, predict potential hazards, and implement preventive measures before incidents occur.

Technology plays a pivotal role in modern contractor risk management. Digital platforms enable real-time compliance tracking, incident reporting, and automated safety training, reducing administrative burdens and enhancing decision-making processes. Companies can also integrate wearable technology and IoT-enabled monitoring systems to track environmental conditions and worker well-being, providing early warnings of hazardous situations.

A comprehensive contractor safety strategy goes beyond regulatory compliance – it fosters a workplace where safety is a shared responsibility, empowering every worker to actively participate in risk prevention. By continuously refining risk assessment methodologies, investing in innovative safety solutions, and fostering strong partnerships with contractors, organizations can establish a resilient safety framework that ensures long-term success in contractor safety management.

REFERENCES

ANSI/ASSP Z10. (2020). *Occupational Health and Safety Management Systems*. American Society of Safety Professionals.

Centers for Disease Control and Prevention (CDC). (2020). *Hierarchy of Controls*. www.cdc.gov/niosh/topics/hierarchy/

CloudSDS. (2024). *EHS Safety Training Update 2024*. CloudSDS. https://cloudsds.com/ehs-update/ehs-safety-training-update-2024/

Cox, T., Griffiths, A., & Rial-González, E. (2000). *Research on Work-related Stress*. European Agency for Safety and Health at Work.

Dekker, S. (2011). *Drift into Failure: From Hunting Broken Components to Understanding Complex Systems*. Ashgate Publishing.

Grote, G. (2012). *Management of Uncertainty: Theory and Application in the Design of Systems and Organizations*. Springer.

Hollnagel, E. (2014). *Safety-I and Safety-II: The Past and Future of Safety Management*. CRC Press.

Hopkins, A. (2019). *Organizing for Safety: How Structure Creates Culture*. CCH Australia Limited.

International Organizations for Standardization (ISO). (2018). *ISO 45001: Occupational Health and Safety Management Systems – Requirements with Guidance for Use*. International Organization for Standardization.

International Organization for Standardization (ISO). (2021). *ISO 45003:2021 – Psychological Health and Safety at Work – Guidelines for Managing Psychosocial Risks*. www.iso.org/standard/64283.html

Leka, S., & Jain, A. (2010). *Health Impact of Psychosocial Hazards at Work: An Overview*. World Health Organization. https://apps.who.int/iris/handle/10665/44428

Manuele, F. A. (2014). *Advanced Safety Management: Focusing on Z10 and Serious Injury Prevention*. Wiley.

National Commission on the BP Deepwater Horizon Oil Spill and Offshore Drilling. (2011). *Deep Water: The Gulf Oil Disaster and the Future of Offshore Drilling*. Government Printing Office.

Occupational Safety and Health Administration (OSHA). (2021). *Recommended Practices for Safety and Health Programs*. www.osha.gov/safety-management

Reason, J. (1997). *Managing the Risks of Organizational Accidents*. Ashgate Publishing.

Safe Work Australia. (2019). *Work-related Psychological Health and Safety: A Systematic Approach to Meeting Your Duties*. www.safeworkaustralia.gov.au/doc/work-related-psychological-health-and-safety-systematic-approach-meeting-your-duties

11 Contractor Emergency Preparedness

Rick L Ingram

11.1 INTRODUCTION

In 1911, future Secretary of Labor Francis Perkins watched from a distance in horror as the Triangle Shirtwaist factory was engulfed in flames. Between smoke inhalation and desperate attempts to keep from burning alive by jumping from the ninth and tenth floors, this infamous fire killed 146 workers, most of whom were young women. Over 25 of the 146 perished after a poorly designed, faulty fire escape collapsed under their weight. As she looked on helplessly, Miss Perkins said, "Something has to be done." From that moment on, Perkins, along with many others, dedicated her career to protecting workers' lives.

This event is only one example of why planning for potential incidents is so important. Many of us have witnessed or heard of other horrible incidents in our lifetimes. An emergency action plan (EAP) can mean the difference between *responding* with confidence to an emergency or *reacting* with panic and confusion, which often results in lives lost. This chapter is meant to provide the basic information that would be helpful for contractors in understanding their legal requirements for creating or improving an EAP. Having a solid plan can certainly prevent Occupational Safety and Health Administration (OSHA) fines and potential legal and civil liability, but even more importantly, it can save lives. I have included several excerpts from the United States Code of Federal Regulations (CFR) in this chapter, but please review all applicable regulations for complete requirements. One last disclaimer – though this chapter is written for and on the basis of the regulatory requirements in the United States, the main principles of a plan for emergencies are basically the same in all country-specific jurisdictions, though some of the specific required components may differ, so use this chapter as a guide and cross-reference your specific local codes if you are outside the United States.

11.2 EAP COMPONENTS

As per OSHA 29 CFR 1910.38, an EAP must include, at a minimum:

- procedures for reporting a fire or other emergency;
- procedures for emergency evacuation, including type of evacuation and exit route assignments;

DOI: 10.1201/9781003679448-11

- procedures to be followed by employees who remain to operate critical plant protocols before they evacuate;
- procedures to account for all employees after evacuation;
- procedures to be followed by employees performing rescue or medical duties; and
- the name or job title of every employee who may be contacted by employees who need more information about the plan or an explanation of their duties under the plan.

As a contractor, you are responsible for your employees' safety even when they work at the hiring client's site. So, you, their employer, must have and maintain an employee alarm system with a distinctive signal for each purpose, in compliance with the requirements in OSHA § 1910.165. In practical terms, it can be something as simple as an air horn, as long as your employees know that one long sound is for fire evacuation, while three short sounds are for weather emergencies, etc. This is just an example of how you can satisfy OSHA requirement with a simple and not very expensive solution, and the specific designations in the previous sentence are just used as an example; you are certainly free to create your own.

As an employer, you also must designate and train your employees to assist in a safe and orderly evacuation of other employees. Additionally, an employer must review the emergency action plan with each employee covered by the plan when the plan is developed, or the employee is assigned initially to a job, when the employee's responsibilities under the plan change, and when the plan is changed.

11.3 STEPS FOR CREATING AN EAP

Ok, but where do I start? you may be asking yourself. Begin by making a thoughtful list of emergencies your company might reasonably encounter based on the work you perform and common natural events at your work locations. Typical action plans include tornadoes, hurricanes, thunderstorms, dust storms, flooding, earthquakes, blizzards, excessive heat, and cold or other potential occurrences.

The next step is to consider personal health emergencies, such as a heart attack, stroke, or serious injuries and illnesses. What response actions would be needed in these situations? Consider both your own internal personnel as well as external responders, such as ambulance, Life Flight, etc.

The next element for consideration should be major mechanical failure or man-made events, such as vehicle crashes, gas leaks, equipment or pipeline damage with or without fire, well control issues, radiological hazards, or chemical spills, to name a few. More recent events, unfortunately, demonstrate a need to include criminal acts, terroristic threats, bomb threats, active shooter(s), and disease or illness outbreaks and responses in your EAP. Think through how you are going to protect your workers and the public, and consider any relevant environmental response as well.

After you have taken the time to brainstorm through all potential emergencies, you are ready to go to the next step, which is figuring out how you should respond in each situation. You briefly thought about this already under the personal medical category, and here you will expand that thought. Among the steps that your employees should take in different types of emergencies are sheltering in place in the event of severe weather, evacuating if there is a fire or tornado, or hiding and barricading during workplace violence situations.

Next, it is imperative that there is a way to easily post, disseminate, or announce emergency numbers and contacts, including fire departments, police, and ambulance services as appropriate. Do not forget to consider any pertinent internal reporting procedures and contacts, including any affected clients or public, as these must also be in the plan.

Then there is a matter of business continuity. As you consider this aspect, think of how you will address all the essential functions, such as IT, accounting, HR, and logistics as applicable.

Finally, your EAP can and really should be as simple as possible. It is *your* plan, so customize it whichever way you need, but keep it simple. In case you are wondering why I mentioned the importance of simplicity twice in a row, it is because the simpler your EAP is, the easier it will be for everyone to understand it, remember it, and practice it. The last part is the most important one and really the reason why you have an EAP in the first place.

As you are developing your EAP, take time to coordinate with your local emergency planning commission (LEPC) and area emergency management coordinator for assistance and guidance, but if not available, contact the local fire and police departments as well as medical providers directly to openly discuss your plan. These key steps will help establish important relationships and ensure success in the event of an actual emergency.

11.4 DEEPER CONSIDERATION OF KEY COMPONENTS

11.4.1 Evacuation Plans

Evacuation plans for buildings must be clearly written and posted, at a minimum by each exit, and must include evacuation routes, location of emergency equipment – such as fire extinguishers, first aid kits, and automatic electronic defibrillators (AEDs) – and muster points or planned employee gathering areas. Although this is not a complicated requirement, many of the evacuation plans that I have examined failed to meet one or more of the criteria mentioned previously.

11.4.2 Exits

Because it is important to ensure an easy and unobstructed path for your evacuating personnel, exits must be:

- clearly marked;
- lighted as needed;
- kept clear and unobstructed at all times;

- wide enough to accommodate the evacuation of employees; and
- not likely to cause additional hazards.

11.5 ALARM/ALERT SYSTEM

A system to alert employees of an emergency, such as fire alarms or all-call systems, must be in place, including a secondary system not dependent on electricity, such as an air horn, as mentioned earlier. In complex plant settings, field locations, or drilling rigs, a system of siren or horn blasts should be developed and reviewed with employees and visitors.

11.6 HANDICAPPED PERSONNEL OPTIONS

Accommodations for the safe evacuation of physically challenged employees or visitors, such as a buddy system, should be incorporated.

11.7 ACCOUNT FOR ALL

Once an evacuation is complete, it must be determined that all personnel are accounted for and safe. This is best accomplished by requiring a sign-in sheet or other method of identification. Employees who are designated to stay behind for specific duties, such as closing valves or other actions, must also be accounted for. One person should be designated to make this accounting and report to management and emergency responders as needed.

11.8 FIRES

Fire emergencies require specific response and equipment. Most businesses depend on the local fire department for any fire beyond the incipient stage. It is up to your management or the host employer to determine whether you should attempt to extinguish a fire or evacuate and call for help. An incipient fire is one that is in the earliest stage, small and controllable with a portable fire extinguisher, standpipe or small hose system, with low amounts of heat and smoke without the need for specialized protective clothing or breathing apparatus.

If fire extinguishers are expected to be used by employees, annual documented hands-on training or the equivalent is required. Extinguishers must be visible, properly mounted, inspected, and maintained by a qualified inspector. Damaged or spent equipment must be removed from service and immediately replaced. Some large, isolated, or complex plants may have in-house fire brigades. When working at a host site, contractors must be familiar with and follow the host company's specific instructions and procedures, but remember that each company is responsible for the safety of its employees.

11.9 MEDICAL

First aid or medical emergencies, like fire emergencies, require specific training and equipment, depending on the location and type of emergency. For general industry, OSHA 29 CFR 1010.151 (b) states that:

> In the absence of an infirmary, clinic, or hospital in near proximity to the workplace which is used for the treatment of all injured employees, a person or persons shall be adequately trained to render first aid. Adequate first aid supplies shall be readily available.

In the construction industry, 29 CFR 1926.50(c) provides:

> In the absence of an infirmary clinic, hospital, or physician, that is reasonably accessible in terms of time and distance to the worksite, which is available for the treatment of injured employees, a person who has a valid certificate in first-aid training from the U.S. Bureau of Mines, the American Red Cross, or equivalent training that can be verified by documentary evidence, shall be available at the worksite to render first aid.

The primary requirement addressed by these standards is that an employer must ensure prompt first aid treatment for injured employees, either by facilitating the availability of a trained first aid provider at the worksite or by ensuring that emergency treatment services are within reasonable proximity of the worksite. The basic purpose of these standards is to ensure that adequate first aid is available in the critical minutes between the occurrence of an injury and the availability of physician or hospital care for the injured employee.

Practically speaking, perhaps the easiest way for the employer to satisfy the regulatory requirements mentioned earlier is by training and certifying a member of your workforce in first aid. For most employers, this is the most feasible and lowest-cost way to protect employees, while ensuring compliance with the standard. If this is what you choose to do, you also need to know that OSHA recommends, but does not require, that every workplace include one or more employees who are trained and certified in first aid, including CPR. As with everything else in this chapter, you will definitely want to check with the host employer or client for their specific requirements, as those may and often do go above and beyond the minimum regulatory requirements.

The other option for employers is to rely upon the reasonable proximity of an infirmary, clinic, or hospital. OSHA has consistently taken the view that the reasonable availability of a trained emergency service provider, such as fire department paramedics or EMS responders, would be equivalent to the "infirmary, clinic, or hospital" specified by the literal wording of the standards. Emergency medical services can be provided either on-site or by evacuating the employee to an off-site facility in cases where that can be done safely. However, the requirements that emergency medical services must be "reasonably accessible" or "in near proximity to the workplace" are stated only in general terms.

An employer who contemplates relying on assistance from outside emergency responders as an alternative to providing a first-aid-trained employee must take a number of factors into account. The employer must take appropriate steps prior to any accident (such as making arrangements with the service provider) to ascertain that emergency medical assistance will be promptly available when an injury occurs. While the standards do not prescribe a number of minutes, OSHA has long interpreted the term "near proximity" to mean that emergency care must be available within no more than 3–4 minutes from the workplace, an interpretation that has been upheld by the Occupational Safety and Health Review Commission and by federal courts. Again, consult with your host employer or client for contractual compliance.

The next thing that needs to be covered is the provision and management of first aid supplies. This is required by 29 CFR 1926.50, which states that first aid supplies shall be easily accessible when required. The contents of the first aid kit shall be placed in a weatherproof container with individual sealed packages for each type of item. Where the eyes or body of any person may be exposed to injurious corrosive materials, suitable facilities for quick drenching or flushing of the eyes and body shall be provided within the work area for immediate emergency use. This covers the provision of supplies.

Given the expirable and disposable nature of the first aid supplies, they must be managed on a regular basis once they have been provided. The employer must check the contents of the first aid kits before sending them out on a job and at least weekly on each job to ensure that the expended items are replaced.

Next, let's talk about transporting an injured employee. The standard requires you, the employer, to have the proper equipment for the prompt transportation of the injured person to a physician or hospital. You are also required to provide a communication system for contacting the necessary ambulance service. This means that in areas where 911 emergency dispatch services are not available, the telephone numbers of the physicians, hospitals, or ambulances shall be conspicuously posted.

In areas where 911 emergency dispatch services are available, and an employer uses a communication system for contacting necessary emergency medical services, the employer must:

- Ensure that the communication system is effective in contacting the emergency medical service.
- When using a communication system in an area that does not automatically supply the caller's latitude and longitude information to the 911 emergency dispatcher, the employer must post in a conspicuous location at the worksite:
  - either the latitude and longitude of the worksite – this is especially relevant to those contractors who work in remote locations or well sites;
  - or other location-identification information that communicates effectively to employees the location of the worksite.

This requirement, specified in 29 CFR 1926.50 (f)(2)(ii)(A), does not apply to worksites with readily available telephone landlines that have 911 emergency service that automatically identifies the location of the caller.

In addition, remote work locations, such as construction or well sites, should have a specific plan, including:

- the latitude and longitude of the site – this could be easily retrieved from most smartphones now;
- driving directions to the nearest medical facility;
- driving directions from the nearest medical facility to the work site;
- emergency numbers;
- and a sign-in sheet for all workers and visitors.

As best practice, the basic EAP elements can be included on the job safety analysis (JSA) form. This ensures that the plan is discussed with and understood by everyone on site and serves as an acknowledgment of review. When planning a new job, visit the worksite to survey potential hazards, and test the emergency numbers to guard against surprises, such as poor cell access or reaching the wrong service. Ask the emergency service providers what the likely response time would be and their preferred procedure to report an actual emergency. Consider the need for an alternate plan for transporting an injured party in places where emergency services are not available. Many smaller service areas have a limited number of ambulances. Consider the possible need for an air ambulance. The information you learn will be invaluable.

For serious injuries – such as those involving stopped breathing, cardiac arrest, or uncontrolled bleeding – first aid treatment must be provided within the first few minutes to avoid permanent medical impairment or death. Accordingly, in workplaces where serious accidents – such as those involving falls, suffocation, electrocution, or amputation – are possible, emergency medical services must be available within 3–4 minutes, especially if there is no employee on the site who is trained to render first aid. This further emphasizes the importance of planning ahead.

OSHA recognizes that a somewhat longer response time of up to 15 minutes may be reasonable in workplaces, such as offices, where the possibility of such serious work-related injuries is more remote. Many companies provide AEDs, which have saved an estimated 1,700 lives in the United States annually. As with other emergency equipment, AEDs must be inspected as per manufacturer recommendations.

Again, keep your plan as simple to follow as possible during an actual emergency. A step-by-step instruction sheet or decision tree for chosen potential emergencies should be at the heart of your written plan. Electronic apps can work, but a notebook doesn't rely on electricity or wireless service to work. This method also serves as an excellent training tool that the employee can reference at any time. Unnecessary contacts and other information should be in a separate section. It is best to keep the primary copy of the EAP in each office, but one with only key response requirements can be created for each vehicle or at remote worksites.

More complicated plans are obviously necessary for major or catastrophic events requiring crisis management teams, complex coordination, etc. However, the main target of this chapter has been a small-to-medium-sized contractor. If you are within that range, a simple plan described here will most likely meet your needs. Remember, at the end of the day, each company must decide what is needed.

As noted earlier, review your EAP with all employees upon hire or when changes to the plan are made, and document all training with the date, summary of covered material, instructor name, and student name. Documented periodic drills and post-drill reviews are key to success, the best measure of effectiveness, and a proven method for continual improvement.

Success in planning for and executing a response to emergencies comes from involving your whole team. Teamwork works. Someone once said: "Failure to plan is planning to fail." Instead of that, I encourage you with the information provided here to have a plan to succeed.

BIBLIOGRAPHY AND FUTURE READING RESOURCES

www.cdc.gov/niosh/docs/2004-101/emrgact/files/emrgact.pdf.

www.govinfo.gov/content/pkg/CFR-2024-title29-vol5/pdf/CFR-2024-title29-vol5-sec1910-38.pdf; www.osha.gov/etools/evacuation-plans-procedures/eap/; www.osha.gov/emergency-preparedness/.

www.nationalstepsnetwork.com/us-hse-toolbox-for-ep, Air Ambulance Protocol.

www.osha.gov/smallbusiness; www.osha.gov/etools/evacuation-plans-procedures/expert-systems/create-eap (build your own EAP electronically).

12 Contractor Assessment Audits

A Strategic Function in Mature Contractor Management Systems

Josh Ortega

12.1 INTRODUCTION

Contractor management systems (CMSs) are foundational to ensuring safe, compliant, and high-performing work environments, particularly in industries where high-risk work is routinely outsourced. As organizations mature in their safety management practices, contractor assessment audits evolve from being merely compliance checkpoints into strategic tools that evaluate contractor capability, culture, and alignment with organizational safety expectations. In a fully developed CMS, audits enable proactive risk identification, performance benchmarking, contractor development, and continuous improvement. This cycle of evaluation and refinement reflects the principles of W. Edwards Deming's Plan – Do – Check – Act (PDCA) model, which emphasizes that improvement depends on learning from performance and using that feedback to take corrective action (Deming, 1986).

This chapter explores the role of contractor assessment audits in mature CMSs, balancing system rigor with behavioral insight. It draws upon international standards (ISO 45001, ISO 45003), systems thinking (Reason, 1997; Hopkins, 2019), risk-based safety management (Manuele, 2014), and behavioral safety science (Geller, 2016), illustrating how audits serve as both control mechanisms and levers for organizational learning.

12.2 AUDITS WITHIN THE MATURITY CURVE OF CONTRACTOR MANAGEMENT

As contractor safety programs evolve, the role of auditing shifts from reactive enforcement to embedded performance enhancement. In immature systems, audits

DOI: 10.1201/9781003679448-12

may simply confirm insurance or checklist compliance. In contrast, mature systems use audits to:

- Validate contractors' real-world application of safety practices
- Assess cultural alignment between host and contractor organizations
- Identify latent risks in procedures, supervision, and communication
- Drive leadership accountability and engagement
- Support ongoing improvement and collaboration

ISO 45001 (2018) provides the structural framework for these systems, emphasizing leadership, participation, and integration. ISO 45003 (2021) further expands this by introducing psychosocial risk as a core safety domain. Audits, therefore, become the interface between these frameworks and field reality.

12.3 STRUCTURING CONTRACTOR AUDITS: SYSTEM RIGOR WITH FLEXIBILITY

Audits in a mature CMS should be structured yet adaptable. Cahill and Kane (2013) describe a practical four-phase model:

1. **Planning** – Define scope, select contractor(s), review scope of work, assign risk tiers.
2. **Execution** – Conduct document reviews, site inspections, observations, and interviews.
3. **Reporting** – Prioritize findings, assign severity levels, and recommend corrective actions.
4. **Follow-up** – Track implementation, verify effectiveness, and close findings.

This model ensures repeatability and defensibility, especially important when contractor incidents are subject to legal scrutiny. However, audits must go beyond templates. Grote (2012) stresses the importance of uncertainty management, recommending that audits assess whether contractors can adapt safely to dynamic conditions, not just adhere to static rules.

A mature audit framework, therefore, blends procedural compliance with real-time adaptability – evaluating how well contractors handle work complexity, shift conditions, and concurrent operations.

12.4 HUMAN AND BEHAVIORAL DIMENSIONS OF CONTRACTOR AUDITING

While structure provides consistency, the human factor perspective is essential to understanding why gaps exist in practice. Reason (1997) emphasizes that major failures often arise not from rule violations but from organizational "latent conditions" – the slow drift of oversight, normalization of deviation, and misaligned incentives.

Audits should thus examine:

- How supervisors reinforce (or ignore) safety norms
- Whether workers feel empowered to raise concerns or stop unsafe work

- The quality of leadership visibility and field-level engagement
- The presence of psychological safety (Hopkins, 2019)

Scott Geller's (2016) behavior-based safety research complements this view by showing that safe behaviors thrive in environments where positive reinforcement, peer accountability, and worker autonomy are present. Instead of using Geller's methods in isolation, mature audits draw from his work to assess the conditions that shape contractor behavior. For example:

- Are safe behaviors recognized and rewarded?
- Are peer interventions encouraged?
- Do contractors implement behavior observation programs?
- Is feedback used to shape habits, not just correct violations?

These questions shift auditing from a backward-looking compliance review to a forward-looking cultural and behavioral assessment.

12.5 LEADING INDICATORS AND RISK-BASED AUDIT FOCUS

Traditional audits often rely heavily on lagging indicators like TRIR, DART, and EMR. However, Hinze et al. (2013) argue for a shift toward leading indicators – metrics that signal future risk rather than record past harm. These include:

- Frequency and quality of safety observations
- Toolbox talk content and participation
- Near-miss reporting and closure rates
- Job hazard analysis (JHA) use and tailoring
- Training completion and verification

Manuele (2014) reinforces this approach by urging organizations to quantify risk exposure, not just incident rates. For contractor audits, this means:

- Evaluating how risk is assessed before starting work
- Observing whether workers understand task-specific hazards
- Auditing how dynamic conditions (e.g., weather, schedule pressure) are factored into decision-making

In mature systems, audits are used to identify which contractors are leading (proactive), managing (compliant), or lagging (reactive) in their risk maturity.

12.6 PSYCHOSOCIAL RISK: A NECESSARY AUDIT DIMENSION

ISO 45003 (2021) introduced psychosocial risk as a formal audit domain. Contractors are particularly exposed due to:

- Temporary work arrangements
- Social isolation on remote or Fly-in, Fly-out (FIFO) projects

- Role ambiguity or conflicting demands
- Limited access to organizational support

Audits in a mature CMS assess whether contractors:

- Identify and control psychosocial risks in their safe work method statements (SWMS) or job hazard analysis (JHA)
- Offer support services (e.g., EAP, critical incident stress resources)
- Train supervisors in mental health awareness
- Promote respectful communication and address harassment

This integration reflects a systems-based, holistic audit approach – evaluating not just physical safety, but psychological safety as well.

12.7 PRACTICAL EXECUTION IN THE FIELD

Mature audits follow clear principles in the field:

- **Triangulation**: Use document reviews, observations, and interviews to verify findings.
- **Focus on real work**: Observe actual tasks – not simulations or prepped examples.
- **Contextual questioning**: Ask "why" and "how," not just "what."
- **Feedback loops**: Deliver real-time coaching and reinforce positive behavior when observed.

Auditors must be trained in communication, behavioral observation, and conflict de-escalation. In high-maturity systems, contractors are invited into the audit process, not just audited upon. Shared accountability and learning replace fear and concealment.

12.8 TECHNOLOGY AND AUDIT INTELLIGENCE

Modern audit platforms enhance maturity by providing:

- *Mobile data collection* tools for real-time capture
- *Audit scoring* based on risk severity and frequency
- *Contractor benchmarking* dashboards across projects or divisions
- *AI-driven trends* that flag repeat nonconformances

Digital tools allow findings to become predictive rather than just historical. When paired with audit narratives and behavioral insights, organizations gain a full-spectrum view of contractor risk and opportunity.

12.9 POSTAUDIT FOLLOW-UP AND CORRECTIVE ACTION CLOSURE

In a mature CMS, the audit process does not end with the final report. Instead, it transitions into the most operationally significant phase: corrective action and closure.

This stage determines whether identified risks are resolved, whether systemic issues are addressed, and ultimately whether the audit has any lasting value. Without effective follow-up, even the most detailed and well-intentioned audits risk becoming symbolic rather than functional (Manuele, 2014).

The primary purpose of audit follow-up is to:

- Ensure that corrective actions have been implemented.
- Validate that those actions are effective in controlling or eliminating the identified hazard or deficiency.
- Track the timeliness of responses, especially for high-risk findings.
- Prevent recurrence by addressing root causes.
- Feed learnings back into the CMS to improve system-wide performance.

Cahill and Kane (2013) argue that the audit process is only complete when every corrective action has been verified and closed through a formal review mechanism This is particularly critical in high-risk sectors (e.g., energy, construction, chemicals), where the failure to close significant audit findings has been a contributing factor in major incidents.

A mature CMS integrates audit follow-up into its broader risk and performance management structure. Common best practices include:

- Risk-Ranked Action Timelines
- Accountability and Ownership
- Verification of Effectiveness
- Closure Documentation
- Escalation Process

ISO 45001:2018 requires that organizations "evaluate the effectiveness of any actions taken" and "make changes to the occupational health and safety management system, if necessary" (Clause 10.2). This mandates a systemic learning response – not just task completion.

Mature systems implement:

- Trend Analysis
- Recurring Issue Flags
- Cross-Contractor Learning
- Linkage to Training and SOPs

Digital contractor platforms and audit systems now automate much of the corrective action workflow, enabling:

- Real-time status dashboards
- Automated reminders
- Integration with HSE risk registers and compliance modules

The value of structured follow-up is clearly demonstrated in the graph given in Fig. 12.1. Over a 12-month period, a major organization implemented a risk-based

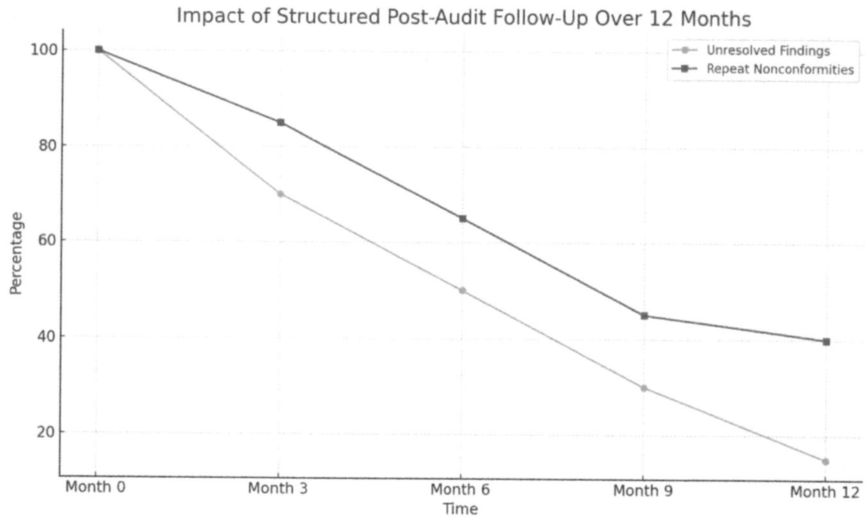

FIGURE 12.1 Impact of Structured Postudit Follow-Up Over 12 Months.

corrective action closure model. Unresolved audit findings dropped by 85%, and repeat nonconformities fell by 60%. These improvements were driven by strong ownership, time-bound action tracking, and leadership engagement in closure verification.

This example illustrates that follow-up is not an administrative burden, it is a strategic intervention. Organizations that prioritize audit closure elevate their contractor management programs from reactive assurance to proactive performance leadership.

12.10 CONCLUSION

Contractor assessment audits, when embedded within a mature CMS, become more than compliance activities – they function as strategic tools for evaluating real-world safety performance, cultural alignment, and risk resilience. Yet the true value of audits lies not only in identifying deficiencies, but also in what follows.

As Deming emphasized, learning and progress occur only when observations inform concrete, system-level change (Deming, 1986).

Postaudit corrective action and closure are essential for translating audit findings into operational improvement. Structured, risk-ranked follow-up processes ensure that identified issues are not only addressed but also resolved in a way that prevents recurrence and strengthens system performance. Organizations that consistently verify the effectiveness of corrective actions, escalate unresolved risks, and communicate lessons learned across teams see measurable improvements in contractor performance and incident reduction.

Research and field data show that structured follow-up dramatically reduces unresolved findings and repeat nonconformities over time. These gains reflect a mature system – one where accountability, learning, and leadership converge to close the loop between audit insight and action.

When supported by strong leadership, behavioral awareness, and risk-based prioritization, contractor audits become a critical driver of safety excellence. They support not just safer work environments but deeper trust with contractor partners and a continuous evolution toward operational excellence and organizational resilience.

REFERENCES

Cahill, R. A., & Kane, R. L. (2013). Environmental Health and Safety Audits (9th ed.). Wiley.

Deming, W. E. (1986). *Out of the Crisis*. MIT Press.

Geller, E. S. (2016). The Psychology of Safety Handbook (2nd ed.). CRC Press.

Grote, G. (2012). Management of Uncertainty: Theory and Application in the Design of Systems and Organizations. Springer.

Hinze, J., Thurman, S., & Wehle, A. (2013). Leading Indicators of Construction Safety Performance. Safety Science, 51(1), 23–28. https://doi.org/10.1016/j.ssci.2012.05.016

Hopkins, A. (2019). Organizing for Safety: How Structure Creates Culture. CCH Australia.

International Organization for Standardization. (2018). ISO 45001: Occupational Health and Safety Management Systems – Requirements with Guidance for Use. www.iso.org/standard/63787.html

International Organization for Standardization. (2021). ISO 45003: Psychological Health and Safety at Work – Guidelines for Managing *P*sychosocial Risks. www.iso.org/standard/64283.html

Manuele, F. A. (2014). Advanced Safety Management: Focusing on Z10 and Serious Injury Prevention (2nd ed.). Wiley.

Reason, J. (1997). Managing the Risks of Organizational Accidents. Ashgate Publishing.

13 Drug and Alcohol Tests for Employees and Contractors

Simon Goncharenko

13.1 INTRODUCTION

Have you ever heard of the Law of Unintended Consequences? The idea behind it is that actions of people or governments more often than not have effects that are unanticipated or unintended.

One example of this phenomenon was the introduction of well-meaning bicycle helmet laws in New Zealand, Australia, and Canada, which focused on increasing the safety of the riders. However, instead of increasing the safety of the riders, these laws had a number of unintended consequences, like:

- suppressing bicycle usage by up to 60%;
- instilling a false sense of security in experienced cyclists, which resulted in faster and harder riding and an uptick in injuries; and
- discouraging female riders, children, seniors, and minorities from even considering riding.

13.2 CONFUSED STATE

It seems that, as of late, we are experiencing this same phenomenon in another area of our lives in the United States – the area of drug use. And it seems that many other parts of the world are very much in the same boat. There seems to be a lot of confusion around the relaxation in medicinal and recreational drug laws throughout the world and all the repercussions that these regulatory decisions spell out for personal lives and work environments.

Take marijuana, for example.

In the United States, federal laws consider marijuana to be a Schedule I drug, with penalties for the possession, distribution, growth, or sale of marijuana. To be clear, President Trump's December 18, 2025 Executive Order on reclassifying marijuana did not change the federal law the moment it was signed. What it did was direct the Attorney General and relevant federal agencies to complete the process of reclassifying marijuana from a Schedule I drug (highest restriction) to a less severe category, like Schedule III, under the Controlled Substances Act. As of the time of this writing,

DOI: 10.1201/9781003679448-13

marijuana remains illegal under federal law as a controlled substance until the drug enforcement administration (DEA) and other agencies take the required legal steps (e.g., regulatory rulemaking) to implement the rescheduling.

Schedule I drugs are also included in the testing requirements for safety-sensitive jobs. However, 92% of the US states, 46 of them to be exact, plus the District of Columbia, have legalized medical marijuana usage, and 48% of US states, or 24 of them, plus the District of Columbia, have legalized recreational marijuana. There are several states, in addition to this, where marijuana is illegal but decriminalized, which means that smoking weed there may result in penalties but not prosecution. And then there are yet others where marijuana is still illegal and not decriminalized, but Cannabidiol (CBD) products are allowed.

So, there is plenty of confusion at the contractor level and that of the individual worker.

Outside the United States, marijuana is legal in Canada, Germany, Mexico, South Africa, and Thailand. Yet even in these geographies, all is not so simple. The European Union illustrates this perfectly. According to John Wicks, cofounder and CEO of Cansford Laboratories and one of the United Kingdom's leading experts in drug testing, workplace drug testing in Europe is a tricky business. Why? Because of the often-conflicting expectations between the privacy laws, the human rights requirements, and the controlled substance regulations, which vary quite widely between different member states.

Likewise, Australia is in flux at the moment, with federal regulations often in conflict with various local, state, and territory laws. When it comes to Asia, unsurprisingly, many countries on this continent maintain strict prohibitionist policies regarding cannabis – the possession, cultivation, trafficking, or use of cannabis is illegal here and can lead to severe legal consequences.

13.3 THE ONUS IS ON THE EMPLOYER

With so much fluctuation in the global regulatory requirements, the onus for implementing workplace controls is on the individual companies. The employers end up being the ones who must both ensure they provide a safe workplace and tread lightly in enforcing their policies and procedures, while dealing with regional privacy laws and medical doctors who may or may not be at liberty to disclose the results of the tests in different jurisdictions.

Then, there is a plethora of special case scenarios, where additional consideration must be provided due to the extenuating circumstances. A good example is those individuals who are approved for medical marijuana but work in safety-sensitive environments, where such use could jeopardize their employment. Additionally, how should the US organizations that are not covered by the Drug-Free Workplace Act function if they still desire to provide a safe work environment for their employees? What is the best way for these companies to protect themselves from the risks of litigation challenging the validity of test results and conflicts with constitutional rights (against unreasonable searches and seizures by the Fourth Amendment of the US Constitution) and other laws (e.g., the Americans with Disabilities Act prohibiting discrimination against qualified individuals with a disability based on prior drug-related problems)? This is a very pertinent question, though answering it may lie outside the scope of the present chapter.

With so many nuances in state/region and federal regulations around the world, the challenge for hiring clients and contractors alike operating within these jurisdictions must be real, especially when it comes to having a clear understanding of what they can and cannot do. This is especially pertinent for those companies that operate in safety-sensitive fields, which may have federally mandated drug testing associated with them.

So, what is the unintended consequence of this confusion between the federal and state requirements?

Individual employees cheating on drug tests.

13.4 WHY EVEN DO DRUG TESTS?

First of all, why even conduct drug tests? From a higher-level perspective, substance involvement represents a significant public and occupational health issue that imposes economic and health-related societal costs. From an employer's perspective, employee drug use can put employers at an elevated risk of accidents, increased absenteeism, lower productivity, and higher insurance costs. Plus, some job families in the United States for example, belong to safety-sensitive categories, which come with an automatic requirement for drug testing.

13.5 IS THE DRUG TESTING SYSTEM PERFECT?

I realize that some will challenge the necessity of drug testing altogether, while others will question the meaningfulness of the metric that is not nuanced enough to test levels of inebriation, like we do with alcohol, while yet others will attempt to prove that failure of a drug test is not causal to an incident. The current system of drug testing in the United States is certainly not perfect, but the presence of an imperfect system is still better than no system at all. Given that the purpose of this chapter is not to defend a particular system and that even an imperfect system can save lives, drug testing is here to stay, at least for the time being.

13.6 WHY TEST FOR MARIJUANA?

Yet others will challenge the necessity of drug testing for marijuana. So, why should we test for cannabis in the workplace? The reasons are quite numerous.

The first reason why we should test for marijuana usage in the workplace is that, quite frankly, nothing good comes from marijuana. Clearly, this statement might seem controversial, as people have argued over the benefits and the drawbacks of drug usage for quite some time now. But the fact of the matter is that there has been no research presenting evidence to date that identifies benefits of marijuana usage to the workplace, while at the same time, there is an overwhelming amount of evidence of the harm and negative outcomes associated with this and other drugs.

Furthermore, the problematic drug consumption creates unique threats to the users and their families' well-being and results in significant economic costs for society by increasing workplace injuries (Spicer et al., 2003), turnover (Bufquin et al., 2021), and absenteeism (Sorge et al., 2020; Van Hasselt et al., 2015). For instance, in 2020, 388 fatal work injury deaths (out of a total of 4,786) were due to unintentional overdose from alcohol and drug consumption, which has quadrupled within a decade (U.S. Bureau of Labor Statistics, 2020, 2021). Here are just a few statistics borrowed from my *Save Lives: Pushing Boundaries in Human Factors* book, where an entire chapter (Chapter 16) is dedicated to this topic:

- Daily cannabis users have a 25% increased risk of a heart attack and a 42% increased risk of a stroke.
- Using marijuana before the age of 18 may permanently affect how the brain builds connections for functions like attention, memory, and learning.
- Marijuana users are twice as likely to get injured on the job when compared to nonusers.
- Of those injured on the job, cannabis users are 22% more likely to end up in ER or be hospitalized.

The list goes on, but the picture is clear. The state laws, which were designed to simplify people's lives and free up the local law enforcement for more weighty matters, ended up pushing people to cheat on the drug tests, increasing the likelihood of getting hurt on the job because of use, exacerbating the extent of injury, and even causing some permanent brain connections which impact the users' attention, memory, and learning.

13.7 WHEN GOVERNMENT RESTRICTIONS EASE, CHEATING ON DRUG TESTS SKYROCKETS

According to Quest Diagnostics' Annual Drug Testing Index, the percentage of US workers whose drug tests showed signs of tampering skyrocketed in 2023 to its highest level in over 30 years. The percentage of employees in the general US workforce whose drug test showed signs of tampering increased by more than sixfold in 2023 compared to the previous year. The increase in substituted urine specimens in the general US workforce was 633%. Invalid urine specimens in the general US workforce increased 45.2%. A result of a substituted or invalid suggests a specimen has been tampered with in an attempt to conceal drug use.

"Given the growing acceptance and use of some drugs, particularly marijuana, it may be unsurprising that some people feel it necessary to try and cheat a drug test," states Suhash Harwani, Ph.D., Senior Director of Science for Workforce Health Solutions at Quest Diagnostics. Dr. Harwani adds that "It is possible that our society's normalization of drug use is fostering environments in which some employees feel it is acceptable to use such drugs without truly understanding the impact they have on workplace safety."

13.8 A WAY FORWARD – WORKERS

From one worker to another worker – my humble advice is don't use marijuana. It's as simple as that. The benefits derived from it are very few and very short-lived, while the negative consequences are very many and most are very permanent.

A recent New York Times article reveals that many marijuana users mistakenly assume that, since the drug is legal in much of the country, it must be nonaddictive and safe. This assumption is dangerously wrong, and many, if not most, of those who begin using the drug for medical purposes report getting addicted and experiencing painful and lasting side effects.

Plus, if you do not use marijuana in the first place, you will never be tempted to cheat on a drug test again.

13.9 A WAY FORWARD – EMPLOYERS, HIRING CLIENTS, CONTRACTORS

As an employer, irrespective of the geography you are in or the specific controlled substance regulatory requirements you are under, you have a legal and a moral obligation to provide a safe and healthy work environment for your employees. As such, it is in your best interest to have sound drug and alcohol and employee assistance policies (EAPs) and procedures to ensure that employee drug testing programs, among other things, are scientifically rigorous and effective.

Let's break this down a bit more here. As a hiring client, you undoubtedly have a sound drug and alcohol policy that outlines the rules and procedures for preventing and addressing substance abuse among your employees. It is imperative that you extend these expectations and requirements to your contractors and their personnel working on your sites and locations. These policies will typically outline requirements for drug testing, including but not limited to:

- pre-employment drug and alcohol testing;
- post-accident drug and alcohol testing;
- random drug and alcohol testing; and
- reasonable suspicion drug and alcohol testing, etc.

Robust drug and alcohol policies for internal employees and contractor personnel will also include awareness programs and disciplinary actions for violations.

Employee assistance programs offer opportunities for employees to seek help for substance abuse issues via resources offered by the employer before policy violations take place.

"Cheating on drug tests not only undermines workplace safety but also jeopardizes the safety of society as a whole," said Katie Mueller, a senior program manager at the National Safety Council focusing on cannabis safety. "Companies, regulators and policymakers must prioritize accountability for the well-being of all individuals in our communities; lives depend on it," she added. Additionally, research suggests that having both a workplace drug and alcohol policy and an EAP is associated with significantly lower rates of marijuana and other illicit drug use compared to having neither of the two measures.

Canadian National Railway is a great example of a company that is successfully navigating the oft-tumultuous regulatory waters for the sake of its workforce, contractors, tenants, and the communities they serve. Their secret? Amidst the confusing and contradictory federal and provincial requirements, Canadian National implemented a robust corporate Policy to Prevent Workplace Alcohol and Drug Problems alongside an Employee and Family Assistance Program. They understand the potentially catastrophic consequences of impairment at work, so they elected to have a zero-tolerance approach to impairment from both legal and illegal substances, not only for their Canadian-based employees but also for their contractors and tenants. Yet at the same time, there is a place for their employees who may be struggling with addictions to seek help before drug or alcohol use becomes a workplace or personal problem.

13.10 A WAY FORWARD – GOVERNMENTS

If you are part of the municipal, state, or federal government that wants to decriminalize or reclassify marijuana to a less stringently regulated controlled substance, you may be responsible for increasing the risk of impaired workers.

How? Because both the decriminalization and reclassification, more often than not, add scores of individuals operating under the influence into the workplaces and the public arena, which could have devastating results.

Take professional drivers, for example. The loosening of regulations could lead to all the professional drivers and transportation workers being untested for marijuana, as Schedule II and III substances are not included in the testing requirements. So, the risk of simply being on the road will increase exponentially, and not just for the drivers, but also for the rest of the public sharing the roads with them.

Please think twice about how your actions may introduce unintended consequences in our workplaces and our communities and cost lives as a result.

13.11 ALCOHOL TESTING

The majority of this chapter focused on the issues related to drug testing, as the question of alcohol testing seems a bit more straightforward and not as complex. In the United States today, safety-sensitive industries have a clear and consistent expectation for testing workers' blood alcohol level in the following circumstances:

- Pre-employment – this is a standard expectation in order to ensure that access to mission-critical and safety-sensitive facilities is reserved for those who are free from the influence of alcohol.
- Random – this is another fairly standard expectation for the employers to be enrolled in some type of a third-party program that randomly tests a certain percentage of their safety-sensitive workers on a regular basis.
- Reasonable suspicion – the management is generally trained on what constitutes reasonable suspicion and have the right to request any employee to be tested during their work hours if the right elements are triggered.
- Post-incident – in order to eliminate the question of whether being under the influence of alcohol or drugs played any role in a workplace incident, this is another standard expectation in American workplaces.

BIBLIOGRAPHY

Bufquin, D., Park, J.-Y., Back, R. M., de Souza Meira, J. V., & Hight, S. K. (2021). Employee Work Status, Mental Health, Substance Use, and Career Turnover Intentions: An Examination of Restaurant Employees during COVID-19. *International Journal of Hospitality Management*, 93, n.pag. https://doi.org/10.1016/j.ijhm.2020.102764.

Frone, M. R., & Bamberger, P. A. (2024). Alcohol and Illicit Drug Involvement in the Workforce and Workplace. In L. E. Tetrick, G. G. Fisher, M. T. Ford, & J. C. Quick (Eds.), *Handbook of Occupational Health Psychology* (3rd ed., pp. 361–383). American Psychological Association. https://doi.org/10.1037/0000331-018

Goncharenko, Simon. (2024). *Save Lives: Pushing Boundaries in Human Factors*. Midway, TX: Kindle Direct Publishing.

Quest Diagnostics. (2024). Workforce Drug Test Cheating Surged in 2023, Finds Quest Diagnostics Drug Testing Index Analysis of Nearly 10 Million Drug Tests. *Newsroom*, May 15. https://newsroom.questdiagnostics.com/2024-05-15-Workforce-Drug-Test-Cheating-Surged-in-2023,-Finds-Quest-Diagnostics-Drug-Testing-Index-Analysis-of-Nearly-10-Million-Drug-Tests

Sorge, J. T., Young, M., Maloney-Hall, B., Sherk, A., Kent, P., Zhao, J., Stockwell, T., Perlova, K., Macdonald, S., & Ferguson, B. (2020). Estimation of the Impacts of Substance Use on Workplace Productivity: A Hybrid Human Capital and Prevalence-Based Approach Applied to Canada. *Quantitative Research*, 111, 202–211.

Spicer, R. S., Miller, T. R., & Smith, G. S. (2003). Worker Substance Use, Workplace Problems and the Risk of Occupational Injury: A Matched Case-control Study. *Journal of Studies Alcohol and Drugs*, 64(4) (July), 570–578. https://doi.org/10.15288/jsa.2003.64.570. PMID: 12921200.

Twohey, Megan, Ivory, Danielle, & Kessler, Carson. (2024). As America's Marijuana Use Grows, So Do the Harms. *The New York Times*, October 4.

U.S. Bureau of Labor Statistics. (2020).

U.S. Bureau of Labor Statistics. (2021).

Van Hasselt, M., Keyes, V., Bray, J., & Miller, T. (2015). Prescription Drug Abuse and Workplace Absenteeism: Evidence from the 2008–2012 National Survey on Drug Use and Health. *Journal of Workplace Behavioral Health*, 30, 379–392. https://doi.org/10.1080/15555240.2015.1047499.

Wicks, John. (2019). Understanding the Workplace Drug Testing Rules around Europe. *Cansford Labs Blog*, January 9. https://blog.cansfordlabs.co.uk/workplace-drug-testing-rules-around-europe

14 Return to Work Reimagined
A Holistic Framework for Employee Well-being and Business Success

Simon Goncharenko

14.1 INTRODUCTION

Despite occupational safety and health improvements in the workplace over the past 30 years, occupational injuries continue to be a prominent public health concern. In 2023, there were over 2.6 million nonfatal workplace injuries and illnesses, according to the US Bureau of Labor Statistics, almost half of which (946,500) involved days away from work (Economic News Release).

Post-injury case management is not a new concept. In fact, it's been around long enough that it's often seen as less important than newer, seemingly more urgent topics like serious injury and fatality (SIF) prevention, human and organizational performance (HOP), or energy-based risk assessments. With the rise of these newer focuses, the retirement of experienced workers, and the influx of younger employees, it's time to bring case management back into the spotlight, as though reintroducing it to the industry for the first time. As we do this, this chapter suggests it may also be time to shift our language from "case management" to "care management."

This subtle but important change better captures the full, shared value of supporting an injured worker's physical and mental health while also responsibly managing the employer's resources. Put simply, updating how we think and speak about this process helps us stay current and serves the best interests of both the worker and the employer. Finally, this chapter is grounded in the belief that work generally supports health by offering purpose, social interaction, and opportunities for growth (Kristman et al., 2020). For that reason, a faster return to work is the central aim of care management.

14.2 LEGACY CASE MANAGEMENT PROCESS OVERVIEW

A large part of today's safety management systems focuses on preventing incidents, injuries, and fatalities, which feels ethically and morally right. Prevention is clearly

DOI: 10.1201/9781003679448-14

important, but we must also recognize that it can never be fully effective (Conklin, Workplace, 54). For prevention to truly succeed, it would need to be tied to our ability to predict these unwanted events. But we simply can't predict exactly what will happen, when it will occur, or how it will unfold. One key reason for this is human fallibility.

People make mistakes – and always will – so there's no way to know where, when, or to whom the next mistake will happen. The only thing we can say with confidence is that mistakes will continue to occur, again and again.

With that in mind, alongside prevention, we must also prioritize recovery – by putting the right controls in place to support safer outcomes. When we design protocols, policies, and standard operating procedures (SOPs) that account for human error, we can better limit the harm caused by incidents we couldn't foresee or stop.

One important recovery method is the holistic management of post-injury care – what we've traditionally called case management. But why is this so important? Workplace injuries and illnesses carry serious consequences. For the injured worker, there's physical pain, mental health struggles, and financial strain from lost or reduced income. For employers, the costs are also steep. They face medical and benefit expenses, lost productivity, rising insurance premiums, higher lagging safety indicators, damage to team morale, reputational harm, and more. In fact, estimates suggest that workplace injuries cost tens or even hundreds of billions of dollars each year (Leigh et al., 2000).

While we can't change what happened, we can influence the process, duration, and ease of recovery – which is where return-to-work programs, or case management, come in. Before exploring effective care management, it's helpful to briefly review traditional case management.

Case management is a systems-based approach to post-injury care, covering everything from the moment an OSHA-recordable injury occurs to full resolution. While definitions vary, here is one of the better ones:

> Case Management is a collaborative process of assessment, facilitation and advocacy of options to assist with an individual's health needs and return to work. Case management is achieved through communication and use of available resources to promote quality and cost-effective outcomes.
>
> (Return to Work Matters)

Serious case management programs involve a cooperative integration between the injured worker, their employer, treating practitioners, external service providers, and claims or insurance companies.

14.3 FROM CASE MANAGEMENT TO CARE MANAGEMENT

The term *case management* feels corporate, impersonal, and overly administrative. It can come across as lacking genuine concern for the individual worker. Often, it seems to prioritize the needs of the employer over the well-being of the employee. Because of these and other negative associations, it's time to rethink and update the language we use.

Care management takes a more holistic approach, aiming for a timely return to work that supports full physical and mental recovery. Research indicates that workers return to work approximately 1.4 times sooner in organizations with care management programs compared to those without them (McLaren et al., 2010). This shortens the median return-to-work time by 3–4 weeks, often benefiting mental recovery even more than physical.

Effective care management understands that injuries impact both the body and the mind. While physical limitations can affect someone's ability to return to work, research shows that mental and emotional challenges after a traumatic injury often play an even bigger role – and may influence recovery more than the physical issues themselves (McCrimmon & Oddy, 2006; Sullivan et al., 2006).

Care management takes a comprehensive approach, balancing physical recovery, mental health, and the employer's supportive role. Effective care management also benefits employers, offering financial, cultural, and reputational gains. Lower worker compensation costs, reduced attrition, and higher employee satisfaction can boost company reputation and attract talent, investment, and contracts.

14.4 THE IMPORTANT COMPONENTS OF GOOD CARE MANAGEMENT

A truly effective care management policy is not developed in a vacuum. It requires close collaboration between the injured worker, the employer, insurance providers, and healthcare providers. While real-world complexities and unique details of various injuries may require more nuanced and targeted implementations, there are key elements that every effective care management strategy should include:

1. Early reporting and prompt response help workers feel comfortable reporting all injuries right away. This will help prevent delays in treatment. To enable early reporting, however, employers must simplify and respond. Employers must simplify the method of reporting and make it intuitive and user-friendly, be it via a 24/7 injury hotline number, digital reporting system, or some other means. Employers must also ensure prompt response to the reporting to further encourage this practice. Moreover, a prompt response will prevent minor injuries from worsening.
2. Utilize professional resources, such as partnering with third-party telephonic case management providers (Corvell, Axiom, WorkCare, etc.). While this is not required, it can be extremely helpful where it is feasible in the process of reaching the aforementioned outcomes. As a pracademic (practitioner and academic), I have both seen the research and experienced the real-world benefits of this approach while leading care management in various organizations. When partnered with a telephonic case management provider, employees are given the provider's number and told to call it first for any nonlife-threatening issue. The telephonic case management partner will triage the call and either provide over-the-phone treatment instructions or set up an appointment at the appropriate occupational clinic. If over-the-phone

treatment is chosen, the telephonic case management provider will follow up the next day and the day after, or whatever intervals are necessary. If a clinic visit is needed, the telephonic case manager shares key details with the clinic to help the doctor understand the injury and its impact on the employee's job. The telephonic case manager keeps the employer updated to ensure everyone stays aligned and works together effectively.

3. Well-defined care management protocols can provide clarity and set everyone's expectations from the start. These protocols may include set workflows for different injuries, clear guidelines for attending medical visits, and predictive analytics tools to spot high-risk cases early.
4. Recognize the impact of organizational culture and provide a supportive environment with strong communication and collaboration. This will help workers feel like they are not alone in this process. The importance of maintaining regular contact between the injured worker, treating physician, and the employer cannot be overestimated. As stated earlier, accompanying the employee to any medical appointments shows care and helps to minimize fraud. Practicing discipline in the careful implementation of this step can also ensure that there is no negative impact on the employee's mental health.
5. Consider the implementation of the stay-at-work initiatives when feasible. This option can support good workplace relationships and speed up recovery by boosting mental well-being. This measure can help spot and address worker issues early, preventing them from getting worse. To keep workers engaged, options include informal accommodations like flexible workloads or rest breaks, adjusting job demands, giving more control or rewards, and offering employee assistance programs. Developing a written "stay-at-work" plan can provide a reference document that all stakeholders can use throughout the process.
6. Focusing on a proactive return to work benefits both the employee and the employer. In his book *Accelerated Recovery*, Dr. Howard B Cotler has a chapter on the science of post-surgery recovery, in which he states that, "In order to return you back to health as soon as possible after your operation, it is important to get you out of bed, walking, eating, and drinking, and off of prescribed narcotics as soon as possible" (Cotler, 40). Dr. Cotler, backed by strong medical research, says that getting back to normal life quickly is the best way to recover faster. It is true in the medical field and should be considered in any holistic care management program. A proactive return-to-work focus helps employees heal faster – physically and mentally – by getting them back to routine sooner. It also benefits employers by improving business operations in several ways.
7. Look for job accommodation options, as finding a solution is typically better than lost time for everyone. Employees stay engaged and recover faster, while employers maintain good relations and some control, which is lost when workers have extended time off. Job accommodation can fill the gaps between the injured worker's temporarily reduced capacities and the job demands. This decision is usually initiated by the employer and reached in collaboration with the injured worker, supervisor, attending physician, and if need be, the insurance providers. In general, job accommodation

needs to be flexible, creative, safe, personalized, temporary, and based on the worker's capacities, workplace challenges, etc. Senior management must support supervisors by giving them flexibility and allowing them to choose the accommodation needed to ensure the success of the return-to-work process (Durand et al., 2014). Senior management support may come in the form of offering more flexible work schedules, temporary reductions in the department's production requirements, etc. Finally, support from coworkers is equally essential for the success of this step.

8. Leveraging industry data in decision-making is greatly underutilized today. While most track key metrics like TRIR, LTIR, and claim costs, many overlook other important data, such as average recovery times for similar injuries. While an internet search can provide the full list, here's a small sample of resources available in the United States:
   - Official Disability Guidelines by MCG Health – mcg.com
   - MDGuidelines by ReedGroup – mdguidelines.com
   - State Workers' Compensation Medical Treatment Guidelines
9. Wellness integration can include many different initiatives, including ergonomic assessments, workplace safety training, human and organizational factors training, promotion of early intervention programs, like stretching and physical therapy, etc.
10. Mental health awareness and integration are closely tied to recovery speed and should be included in effective care management. While the connection between care management and mental health is not usually at the forefront of the safety professional's mind, the implications must not be neglected. There is usually a strong correlation between the length of recovery and the negative impact on one's mental health. Sadly, most injured workers may not realize this, and many care management programs don't consider it. Elizabeth Van Horn's 2009 study on loss and depression after traumatic injury highlights key factors impacting mental health, including loss of independence, unachieved goals, decreased personal success, feeling less valuable, loss of pride, and reduced control over life. These factors lower self-esteem and can lead to increased anxiety and depressive symptoms (Van Horn, 2009).

14.5 THE IMPACT OF NONEXISTENT OR INEFFECTIVE CARE MANAGEMENT

The impacts of nonexistent or ineffective care management are significant. According to a 2016 study, participants from employers with nonexistent or ineffective care management were 23.22 times less likely to return to work than those from employers with effective care management (Awang et al., 2016). Additionally, participants whose intervention period exceeded 5 months were 41% less likely to return to work compared with those whose intervention period was within 3 months.

In fact, it can be argued that the impact and cost of ineffective care management is much greater to an organization than the cost of the initial injury. While not all incidents can be prevented, good care management is 100% within the organization's ability to deliver. There is no excuse for failing to prepare, execute, and follow through consistently and effectively in the area of care management.

What are the actual costs of ineffective care management? First, there is the actual cost of medical care, which can be much higher if all employees are left to utilize the local emergency room as the default place of care. Second, without an effective care management program, there's a higher risk of fraud or misuse by both the injured worker and the medical provider. Third, there is the increased cost of worker compensation insurance when the employer takes a lackadaisical approach to care management.

Fourth, a reactive approach to care can raise the employer's experience modification rate (EMR), a 3-year risk rating used by insurers for workers' comp coverage. The higher the EMR, the costlier the insurance premiums. An EMR of 1.0 is average – lower EMR indicates a safer workplace and lower premiums, while higher EMR indicates more risk and higher costs. Additionally, an EMR of greater than 1.0 can result in more difficulty with qualifying for lucrative contracts and/or outright rejection from the bidding pool.

Lastly, poor care management can harm an employee's mental health, making them feel ignored and unimportant to their employer. At the same time, they may be targeted by legal ads encouraging them to take full advantage of the system. The biggest problem with this situation is that the only winners are the attorneys, while the injured workers and the employers come out on the losing side. The injured workers become depressed, have a tarnished reputation, and often end up losing their jobs, which can result in income interruption and impact relationships at home. Employers end up in costly legal battles they can't win, hurting profits and lowering morale among other workers.

14.6 HOW EFFECTIVE CARE MANAGEMENT IS IN EVERYBODY'S BEST INTERESTS

Effective care management, on the other hand, is truly in the best interest of all parties involved. The injured worker does not have to handle all the repercussions of their injury alone. There is a clear and well-defined process in place that guides them on what to do, what to expect, and what resources they can use. When the employer helps the injured worker to get to the clinic, accompanies the individual during their care, and is otherwise fully engaged, it helps the individual to recover faster and allows them to get back to work sooner. Feeling engaged speeds up their physical recovery while also supporting their mental health. What's often missed in past research is that a workplace injury can impact not just the injured worker's mental health, but their teammates' too. Good care management can help support the mental well-being of the whole organization.

It goes without saying that effective care management is also good for the employer. The frequency and severity of the injuries tend to go down. The willingness to report injuries tends to go up. The cost of injuries greatly decreases. This in turn decreases the cost of insurance and brings down the EMR, which results in more contracts, better opportunities to bid, and other reputational advantages.

14.7 INDUSTRY EXAMPLE OF TANGIBLE RESULTS FROM EFFECTIVE CARE MANAGEMENT

Leveraging strong care management and safety leadership to handle both short- and long-term conditions helps reduce injury severity and creates a more productive workplace (Daniel & Albert, 2016). ExxonMobil Global Services Company (GCS), a subsidiary of ExxonMobil Corporation, employing 9,000 workers and over 12,000 full-time equivalent contracted personnel, discovered that a switch to more holistic care management resulted in employees no longer being hesitant to report injuries. Additionally, workers came forward earlier, particularly in chronic musculoskeletal cases, before their conditions became severe or debilitating. This shift in approach was a significant contributor to minimizing the total number of OSHA-defined recordable injuries by 65% between 2010 and 2014. During this same period, the total number of injury incidents reported by workers actually increased by 20%.

14.8 CONCLUSION

In summary, this chapter reintroduced the concept of return-to-work, or case management, arguing for its equal benefit for the employee and the employer alike. This chapter also argued that it may be time for the upgraded label from case management to care management to fully reflect the overall approach encapsulated in this area. Finally, mental health is often more impacted than physical health by workplace injuries, and a comprehensive care management approach can make a big difference.

BIBLIOGRAPHY

Awang, Halimah, Shahabudin, Sharifah Muhairah, & Mansor, Norma. (2016). Return-to-Work Program for Injured Workers: Factors of Successful Return to Employment. *Asia Pacific Journal of Public Health*, 28(8), 694–702.

Conklin, Todd. (2017). *Workplace Fatalities: Failure to Predict: A New Safety Discussion on Fatality and Serious Event Reduction*. Santa Fe, NM: CreateSpace Independent Publishing Platform.

Cotler, Howard B. (2016). *Accelerated Recovery of Your Health: How to Recover Your Body After Injury or Surgery*. Ocala, FL: Atlantic Publishing Group, Inc.

Daniel, M. Couch, & Albert, J. Robb. (2016). *Effective Care Management Minimizes Workplace Injury Severity*. Paper presented at the SPE International Conference and Exhibition on Health, Safety, Security, Environment, and Social Responsibility, Stavanger, Norway, April 2016 (SPE-179229-MS). https://doi.org/10.2118/179229-MS

Durand, Marie-Jose, Corbiere, Marc, Coutu, Marie-France, Reinharz, Daniel, & Albert, Valerie. (2014). A Review of Best Work-absence Management and Return-to-work Practices for Workers with Musculoskeletal or Common Mental Disorders. *Work*, 48(4), 579–589.

Employer-Reported Workplace Injuries and Illnesses, 2023. (2024). *Economic News Release*. www.bls.gov/news.release/osh.nr0.htm

Goncharenko, Simon. (2024). *Save Lives: Pushing Boundaries in Human Factors*. Midway, TX: Kindle Press.

Goncharenko, Simon. (2025, forthcoming). *Operationalizing Twenty First Century Safety: A Humancentric Practical Guide*. Boca Raton, FL: CRC Press.

Kristman, V. L., Boot, C. R. L., Sanderson, K., Sinden, K. E., & Williams-Whitt, K. (2020). Implementing Best Practice Models of Return to Work. In U. Bültmann & J. Siegrist (Eds.), *Handbook of Disability, Work and Health. Handbook Series in Occupational Health Sciences* (Vol. 1). Cham: Springer. https://doi.org/10.1007/978-3-319-75381-2_32-1.

Leigh, J. P., Markowitz, S., Fahs, M., & Landrigan, P. (2000). *Costs of Occupational Injuries and Illnesses*. The University of Michigan Press.

McCrimmon, S., & Oddy, M. (2006). Return to Work Following Moderate-to-Severe Traumatic Brain Injury. *Brain Injury*, 20, 1037–1046.

McLaren, Christopher F., Reville, Robert T., & Seabury, Seth A. (2010). *How Effective Are Employer Return to Work Programs?* RAND Institute for Civile Justice Working Paper Prepared for the Commission on Health and Safety and Workers' Compensation.

Sullivan, M. J. L., Adams, H., Thibault, P., Corbiere, M., & Stanish, W. D. (2006). Initial Depression Severity and the Trajectory of Recovery Following Cognitive – Behavioral Intervention for Work Disability. *Journal of Occupational Rehabilitation*, 16, 63–74.

Van Horn, Elizabeth. (2009). Loss and Depression After Traumatic Injury: The Importance of Case Management in the Recovery Process. *Professional Case Management*, 14(2), 66–73.

15 Incident and Near-Miss Investigations You Should Expect from Your Contractors

Karl DeLooff

15.1 INTRODUCTION

Effective incident and near-miss investigations are foundational to a resilient industrial safety management system. Drawing on the combined insights from US Coast Guard investigative doctrine, OSHA guidelines, the National Safety Council (NSC), and personal experience, this chapter articulates a comprehensive approach that emphasizes system thinking, organizational accountability, and preventive action. This chapter integrates James Reason's seminal work on human error and latent organizational failures to support a just and learning-focused investigative culture.

15.2 PURPOSE OF INCIDENT INVESTIGATION

Incident investigations aim to prevent recurrence, not assign blame. As OSHA (2015) emphasizes, many workplace incidents can be prevented when the underlying programmatic or systemic failures are addressed. The same principle underpins the leadership principles, which emphasize that investigations should determine root causes and establish corrective actions that eliminate system failures and account for unavoidable human errors and factors.

The purpose is not merely to answer, "what happened?" but to systematically answer:

- What happened?
- How did it happen?
- Why did it happen?
- What can be done to prevent it from happening again?
- How can the negative consequences of incidents be minimized?

DOI: 10.1201/9781003679448-15

15.3 SCOPE AND CLASSIFICATION OF INCIDENTS

An incident, as defined by the leadership and management procedure, is an unplanned, undesired event that adversely affects task completion. Incidents include but are not limited to:

- Significant near misses
- Recordable injuries
- Property damage
- First aid and medical treatment cases
- Detected pathway errors

All incidents must be reported and appropriately investigated, with the depth of investigation guided by the potential and actual severity. Near-miss incidents, often underreported, offer critical insight into latent system vulnerabilities and are emphasized in both OSHA and NSC literature as essential learning opportunities.

Pathway errors, where an incident could have occurred, but was stopped by a defense, should also be investigated. The intended outcome is an understanding of the error or failure mode for corrective or improvement actions and the strengths and weaknesses of the defense. Potential defense weaknesses are a main source of latent safety and quality issues that only become manifest under rare circumstances that can have significant consequences. In complex cases, a critical path exists which, if interrupted by defenses, would have prevented the incident or significantly reduced its negative outcomes.

15.4 GUIDING PRINCIPLES OF INVESTIGATION

Drawing on James Reason's work (1990, 2000), modern incident investigation transcends the blame model and instead probes systemic weaknesses. The following principles apply:

- **Just Culture**: Create an environment where individuals feel safe reporting mistakes.
- **Systems Thinking**: Errors are outcomes, not causes. Investigations should examine organizational, environmental, and latent conditions.
- **Objectivity and Facts**: Investigate to uncover facts, not assign personal fault (NSC, 2014).
- **Root Cause Focus**: Ask "why" until the fundamental cause or a pathway is identified. Many incidents will have a critical path rather than a single root cause.

15.5 INVESTIGATION PROCESS OVERVIEW

Following the OSHA four-step system (OSHA, 2015) and most other models, the process consists of:

15.6 STEP 1: IMMEDIATE RESPONSE AND SCENE CONTROL

- Render aid and stabilize the scene.
- Secure the area to preserve evidence
- Notify relevant leadership per protocol (within 2 hours per most company procedures).
- Document initial facts and initiate witness separation.

15.7 STEP 2: EVIDENCE COLLECTION AND INFORMATION GATHERING

- Document the scene via photographs, sketches, and evidence tagging.
- Interview witnesses using open-ended, fact-focused questions.
- Collect documentation: job safety analyses (JSAs), permits, training records, maintenance logs. Due to the Internet of Things (IoT) and supervisory control and data acquisition (SCADA) systems, immediately preserve the electronic and digital records. Immediately take photographs to preserve analog conditions, including gauge readings, temperature, placement, and other factors that may have a limited life or duration.
- Maintain a neutral, nonpunitive tone to encourage honesty (NSC, 2014).

15.8 STEP 3: ROOT CAUSE ANALYSIS

- Develop an event log/timeline for the incident. This log should also include the systems, processes, and equipment to provide context. By extending the log and including the context, latent factors are potentially discovered, leading to corrective or improvement actions.
- Utilize structured tools such as:
  - Five Whys (or other protocols such as TapRoot, etc.)
  - Fishbone/Ishikawa diagrams
  - Event and causal factor charting (OSHA, 2016)
- Involve subject matter experts and frontline personnel. One of the main concerns is confirmation bias – having a theory that leads to selective evidence. The primary way to avoid confirmation bias is to acknowledge it and have a disciplined procedure that all personnel will agree to follow.
- Examine contributing factors in the domains of people, procedures, equipment, environment, and organizational oversight.

15.9 STEP 4: CORRECTIVE ACTION AND FOLLOW-UP

- Develop corrective and preventive actions (CAPAs) that target root causes.
- Validate actions through management systems or equivalent oversight (per internal procedure).
- Track completion in a central system (e.g., Enablon, Intelex, or similar database software).
- Monitor for effectiveness post-implementation.

15.10 OPERATIONALIZATION

- Investigative teams should incorporate a "capability–condition mismatch matrix" (see the following section) as part of evidence analysis. A capability–condition mismatch matrix is a risk assessment and planning tool used to identify situations where there is a gap (mismatch) between the capabilities available (people, equipment, plans, procedures) and the conditions faced (hazards, demands, environments).
- Questions from the US Coast Guard Maine Safety Manual (USCG MSM) include: "Was the asset design or configuration suitable for this mission?" or "Were personnel briefed with timely and accurate information?"
- Discrepancies should be documented and correlated with contributing factors and latent conditions.

15.11 CAPABILITY–CONDITION MISMATCH MATRIX AND SECTOR-SPECIFIC EXAMPLES

The following table outlines common mismatches observed in wind turbine, Photovoltaic (PV) solar, and electrical interconnect substation operations, along with examples from real-world investigations.

TABLE 15.1
Common Mismatches in Wind, Solar, and Substation Operations

Sector	Condition	Capability Mismatch	Example/Reference
Wind Turbine Operations	Emergency descent during high wind	Inadequate training in descent systems	2013 Bakken Energy Services incident – two fatalities due to poor emergency egress competency (OSHA).
Wind Turbine Maintenance	Hydraulic system lock-out	Insufficient procedural clarity	OSHA 2021: Technician crushed due to failure to follow unclear hydraulic lockout protocol.
PV Solar Farm Installation	Working at height on tilted racks	Lack of fall protection awareness	NSC (2014): Near-miss due to misused harness, no rescue plan established.
PV Solar Farm O&M	Arc flash during inverter service	No arc flash hazard assessment or PPE	OSHA 3895 (2016): Burn injuries resulted from unmitigated arc flash hazard exposure.
Substation Switching	Manual switching under load	Inadequate electrical training	2017 NIOSH Fatality Report: Contractor electrocuted while switching under live load.
Substation Commissioning	Temporary grounding layout	Technicians unfamiliar with the setup	Based on 2019 Energy Safety Canada review: Grounding omitted during commissioning led to incident.

In each of these cases, a gap between operational conditions and the available capability – training, procedural access, proper tools, or asset configuration – contributed to the incident. These examples reinforce the need to use the MSM framework proactively during investigations to surface latent mismatches before they result in harm.

In addition, the US Coast Guard Marine Safety Manual (MSM), Volume V on Investigations, offers a valuable framework for identifying mismatches between operating conditions and the capabilities of personnel, resources, assets, or information. When conducting root cause analysis, investigators should consider the following mismatches as part of systemic evaluation:

- **Capability Gaps**: Was the individual or team adequately trained or experienced for the task?
- **Resource Deficiencies**: Were tools, equipment, or protective systems lacking or inadequate?
- **Asset Suitability**: Were the physical assets (e.g., vehicles, vessels, machinery) appropriate for the environment and the task?
- **Information Availability**: Were instructions, procedures, or hazard communications accessible and understood by the workforce?

This systemic check against capability mismatches aligns with Reason's (1997) model of latent conditions and sharp-end actions – helping ensure investigators examine whether the system equipped people for success.

15.12 INTERVIEWING SKILLS

The primary objective of an interview during an industrial incident investigation is to uncover the truth, discover relevant facts, and provide contextual understanding of the event. An effective interview allows the witness or subject to tell their story in their own words, ideally in the order they experienced it. Investigators should aim to facilitate a natural, uninterrupted narrative, as individuals typically recall events in fragmented "packets" of information tied to specific triggers or emotions.

Interviewees often compress time or relay events out of sequence, making it essential to record their account verbatim and later reconcile it with the established event timeline. During this process, investigators may need to revisit specific points to clarify the proper chronological placement.

To encourage detailed and continuous narration, several interviewing techniques have proven highly effective. Among these are mirroring, labeling, and elicitation, as popularized by Chris Voss in *Never Split the Difference* (Voss & Raz, 2016) and by Chase Hughes in the *Behavioral Science Unit's Behavioral Analysis Manual* (Hughes, 2017).

- **Mirroring**

Mirroring involves repeating the last two to three words the interviewee has just said, or the keywords from a sentence the investigator wishes to explore further. This subtle prompt signals active listening and encourages the subject to continue speaking. Because most people recount incidents in segments, mirroring helps maintain the flow of conversation, allowing a more complete picture to emerge.

- **Labeling Emotional Content**

Labeling is used to identify and name the emotions conveyed during the narrative. Most individuals involved in incidents experience emotional responses such as anxiety, guilt, regret, or even excitement. Investigators can prompt deeper reflection by stating, for example, "It seems like that was frustrating" or "It sounds like you were concerned." Allowing the label to hang in silence typically invites the subject to elaborate, providing greater emotional context for their actions or perceptions during the event. This method is derived from both Voss's negotiation techniques and the reflective practices developed by psychologist Carl Rogers.

- **Elicitation**

Elicitation, as outlined by Hughes, involves making a seemingly factual statement – correct or incorrect – to prompt correction or clarification. Rather than asking direct questions, which may cause defensiveness, this technique leverages human tendencies to correct misinformation. Delivered with humility and curiosity, elicitation is a powerful way to draw out nuanced information that may not surface through standard questioning. For example, "Everybody knows that process won't work." This could elicit the response of: "But we made it work via this and this hack." Or providing a high and low estimate in order to entice a more specific number: "I'd guess between 5 and 15 dollars." This could elicit the response of: "Probably around seven dollars."

- **Constructing the Event Timeline**

A core goal of the interview process is to develop an event log or timeline, which must later be cross-validated with physical evidence and digital records (e.g., SCADA data, alarm logs). Investigators should place greater confidence in objective data, as human memory is often distorted by stress, emotion, or sequential inaccuracies. Individuals may perceive several events occurring concurrently even when digital evidence records them as distinct, time-stamped actions.

Once an event log is drafted, the investigator should review it with the interviewee, confirming the sequence of events – not necessarily the exact timestamps. It's acceptable and common for multiple actions to occur within the same short window (e.g., 2–4 events within 5 seconds), and this needs to be acknowledged in the report. SCADA and digital systems often provide millisecond resolution, while human perception clusters events as happening simultaneously.

- **Reverse-Order Recall**

One method for assessing the reliability of an account is reverse-order recall. Instead of progressing forward through the timeline, the interviewer prompts the subject to describe the sequence in reverse: starting from the outcome and working backward through preceding events. This technique accesses different cognitive pathways and often bypasses rehearsed or socially filtered narratives, leading to sharper, more accurate recollections.

- **Probing for Tribal Knowledge and Deviations**

Investigators must also probe for tribal knowledge – the informal, often undocumented methods workers use to accomplish tasks. These adaptations may differ significantly from standard operating procedures and are critical for identifying systemic vulnerabilities or gaps in formal training. Recognizing where individuals deviate from procedures, and why, provides insight into operational drift and latent organizational weaknesses.

- **Obstacles, Objections, and Frame of Reference**

Key areas of exploration during the interview include:

- Obstacles: External factors that hindered the subject's ability to perform a task (e.g., equipment malfunction, environmental hazards).
- Objections: Internal resistance or ethical concerns that led to deviation from expected behavior.
- Frame of Reference: The subject's mindset and expectations at different stages – assignment, planning, execution, incident occurrence, and aftermath. These can shift rapidly during dynamic operations and shape decision-making under pressure.

- **Documentation: Separating Facts from Analysis**

Interview documentation must clearly distinguish between objective facts and subjective analysis. The event log and transcript should reflect the witness's account without interpretation. In a separate section, the investigator may present their analysis and findings derived from the interview, explicitly noting where interpretations were made, and which reasoning supported those conclusions.

A powerful tool in this context is restatement, wherein the investigator summarizes the subject's statements and seeks affirmation – aiming for the subject to respond with, "That's right," which signifies mutual understanding. This confirmation indicates not only agreement but also that the investigator has captured the true intent behind the subject's account.

Observations of nonverbal behavior, demeanor, and stress indicators during the interview also contribute valuable context and should be included in the investigator's notes, provided they are presented with clarity and without bias.

15.13 THE ROLE OF LEADERSHIP AND CULTURE

Leadership's role is critical in setting an investigative tone and supporting systemic improvements. Company policy and practice should mandate the involvement of managers, safety professionals, and site leadership in determining investigation level and participating in fact-finding. Leaders must also model openness to learning and support transparent root cause discussions.

James Reason (1997) emphasizes that organizational accidents arise from latent conditions that lie dormant within the system until triggered by active failures. Effective leadership is proactive in identifying and eliminating these latent hazards.

Leadership owns the system, policy, and procedures. Under the concept of *Extreme Ownership* (a book published in 2015 under the same title by Jocko Willink and Leif Babin), no progress or improvement is possible without taking ownership of the situation and the potential solutions. Latent conditions, by nature, are owned by the company's leadership or by another authority or decision-maker. Consequently, leadership's involvement in the investigation process is not to assign blame, but rather to assign responsibility for corrective or improvement actions. In some cases, there may be preventative actions that become part of the defensive systems and barriers to future incidents.

15.14 COMMON PITFALLS AND HOW TO AVOID THEM

- **Blame Culture**: Leads to fear and incomplete reporting. Instead of blaming, promote psychological safety. While there may be personal decisions that might have led to the incident that may require discipline, the overall culture should be one of justice.
- **Stopping at Human Error**: Treating human error as a cause rather than a symptom precludes deeper learning (Reason, 2000).
- **Superficial Corrective Actions**: Addressing only proximate causes (e.g., "retraining") without system change fails to prevent recurrence. It is often much simpler to try to fix the worker rather than the worker's environment and controlling systems. Systems must be placed to result in positive outcomes, not create inevitable errors.
- **Inconsistent Investigations**: Use standardized tools and protocols (OSHA, NSC, and internal). Investigator discipline and consistency are part of the culture that ensures the culture of justice.

15.15 SPECIAL CONSIDERATIONS FOR NEAR-MISS INVESTIGATIONS

Near-miss events, particularly those classified as significant, should receive the same investigative rigor as actual incidents. As Reason (2008) and OSHA (2015) underscore, near-misses are gifts – low-cost opportunities to uncover systemic weaknesses.

- Treat near-miss reporting as proactive risk identification.
- Conduct full investigations into significant near-misses (as defined by company protocols).
- Encourage open reporting through positive reinforcement and anonymity options. Have multiple channels for reporting. What is accessible and easy is more likely to be performed.

15.16 INTEGRATING FINDINGS INTO ORGANIZATIONAL LEARNING

The ultimate goal of investigations is organizational learning. This requires:

- Documenting findings in an accessible format.
- Sharing lessons learned across similar operations.
- Tracking trends and identifying systemic vulnerabilities.
- Incorporating findings into training, design, and policy updates.
- Applying incident and outcome-based reviews to confirm that systemic capability mismatches are not repeated.
- Socializing the findings and systemic improvements for incidents, near-misses, and pathway errors.
- Creating a culture of learning and curiosity.

15.17 CONCLUSION

A robust incident and near-miss investigation process transforms reactive risk management into proactive hazard control. By integrating the systems thinking advocated by James Reason, the structured methods outlined by OSHA and NSC, and the mismatch detection techniques from the USCG MSM, organizations can significantly reduce recurrence, strengthen culture, and capacity to recover, and advance toward safety excellence.

BIBLIOGRAPHY AND FURTHER READING

Energy Safety Canada. (2019). *Electrical Safety Program Review.*

Hughes, C. (2017). *The Ellipsis Manual: Analysis and Engineering of Human Behavior*. Evergreen Press.

National Safety Council. (2014). *How to Conduct an Incident Investigation.*

NIOSH Fatality Investigation Reports. (2017). U.S. Centers for Disease Control and Prevention.

OSHA. (2015). *Incident [Accident] Investigations: A Guide for Employers*. U.S. Department of Labor.

OSHA. (2016). *Root Cause Analysis Fact Sheet* (Publication No. OSHA 3895). U.S. Department of Labor.

Reason, J. (1990). *Human Error*. Cambridge University Press.

Reason, J. (1997). *Managing the Risks of Organizational Accidents*. Ashgate.

Reason, J. (2000). Understanding Adverse Events: Human Factors. *Quality in Health Care*, 9(suppl 2), ii20–ii25.

Reason, J. (2008). *The Human Contribution: Unsafe Acts, Accidents and Heroic Recoveries*. Ashgate.

Rogers, C. R. (1951). *Client-Centered Therapy: Its Current Practice, Implications, and Theory.* Houghton Mifflin.

U.S. Coast Guard. (2021). *Marine Safety Manual, Volume V – Investigations*. COMDTINST M16000.10(series).

Voss, C., & Raz, T. (2016). *Never Split the Difference: Negotiating As If Your Life Depended On It*. Harper Business.

16 When Low Injury Rates Lie

The Hidden Face of SIFs

Simon Goncharenko

16.1 INTRODUCTION

One of the topics that is quickly gaining momentum in the world of safety is regarding how to break the stalemate on serious injuries and fatalities (SIFs). The reason for this focus is simple. Over the past 30+ years, here in the United States, we have successfully decreased the overall OSHA recordable rates by over 60%. That is impressive (even if the total recordable incident rate (TRIR) is not a helpful indicator).

16.2 THE CURRENT SIF STALEMATE

The progress has not been the same with SIFs, however. So, while we may be doing quite well in decreasing the rates of twisted ankles and sprained knees, the stuff that kills and maims our employees has been a little more difficult to control at a similar pace. In fact, it is fair to say that we have stalled out in our effort to decrease the SIF rates. Some of the latest research actually suggests that the number of SIFs has not only refused to budge, but it has also been on the rise.

16.3 WHAT CAUSES SIFS IS DIFFERENT

Why? Because, first of all, the things that lead to recordable injuries and the things that prevent or control them are often markedly different from the things that lead to SIFs. And when we focus on preventing strains and sprains, our efforts may not result in decreasing the more significant injuries as well.

16.4 PLAYING THE TRIR GAME

Why else? Because if we are honest, many of us have learned how to play the TRIR game well enough to keep many incidents from being recordable. Sidney Dekker calls this game Safety Theater (Dekker, "Safety Theater," *PSJ*). Some of you may be getting offended by now because TRIR is all you know, and you are quite comfortable operating within this arena.

DOI: 10.1201/9781003679448-16

16.5 LIMITATIONS OF RECORDABLE RATE METRICS

To the readers who are still basing their entire approach to safety on the TRIR/TCIR rates and the like, let me offer some food for thought. I am not trying to talk you out of anything or into anything. I am merely ensuring that you are sufficiently educated on the limitations of the recordable rate metrics.

First, from a purely scientific perspective, TRIR/TCIR is not statistically stable. According to a 2018 case study performed by Kotek, Nosek, and Bartos, a metric is statistically valid if it is based on sufficient volumes of consistent data within reasonable time frames to allow acceptable uncertainty estimations and statistical precision (Safety Metrics of Performance for Small and Medium-Sized Enterprises). In layman's terms, for your organizational TRIR/TCIR to carry statistical meaning, you will have to accumulate a billion-plus worker hours. Dr. Matt Hallowell from the University of Colorado did an evaluation in 2021 of more than 3 trillion worker hours and found that the occurrence of recordable injuries is almost entirely random (The Statistical Invalidity of TRIR as a Measure of Safety Performance).

So, next time you are looking at your TRIR/TCIR, ask yourself if it was calculated on the basis of a billion-plus worker hours. If not – and let's be honest, most of them are not – then this metric has no statistical meaning in your organizational story.

But there is more that is wrong with looking solely at TRIR/TCIR. The second thing that this lagging indicator does not possess is predictive capacity. TRIR/TCIR of the past is not indicative of the TRIR/TCIR in the future.

The third reason why lagging indicators alone are an insufficient foundation for a good, solid safety program is because of the implied definition of safety that they operate under. TRIR/TCIR and other lagging metrics assume that safety is the absence of injuries. However, anyone who has done a deeper dive into how actual work is done understands that some days, jobs, and projects may escape an injury not due to the safe way of working but due to nothing but dumb luck, as my seasoned construction superintendent friend Peter Johnson puts it.

So, your focus on the recordable metrics and the Safety Theater performance may have enabled you to decrease your company's TRIR, but the story is not the same with SIFs. And that is why so many of the biggest and most forward-thinking organizations today, like Amazon, Boeing, Tesla, Home Depot, American Airlines, Delta Airlines, GM, Disney, Lockheed Martin, Google, AECOM, and others are beginning to track SIFs while continuing to pay attention to the other OSHA recordables as well.

16.6 COMMON DEFINITION OF SIFS

Of course, one of the important first steps in this approach would be to arrive at a definition of SIF that everyone can agree on. Without a common definition, there cannot be a unified effort at defeating these injuries.

Amidst numerous attempts to define "serious," one that seems to be the most straightforward considers how the injury impacts a worker's life. According to this way of thinking, if the injury is life-ending, life-threatening, or life-altering, it qualifies as a SIF. For a more detailed discussion of this approach, see "What is a Serious Injury?" in the September 2023 issue of *Professional Safety Journal.* You can also check out the research conducted by the Construction Safety Research Alliance at the University of Colorado – www.csra.colorado.edu. This organization focuses on the construction industry, but its findings apply to all industries.

16.7 HOLISTIC STRATEGY OF TRACKING TRIR AND SIFS

When you visit their website, make sure you read about their efforts to replace TRIR with high energy control assessment (HECA) rate as a more reliable and statistically valid method to eliminate SIFs. HECA measures the percentage of high-energy hazards with direct control in place, which allows you to see the effectiveness of that control. Their proposed formula for HECA measurement is Success ÷ (Success + Exposure). Their other proposed metric measures the HECA activity rate, which normalizes the volume of high-energy control verifications to 100 full-time employees. HECA activity rate equals to (HECA Assessments × 200,000) ÷ Hours Worked.

Another proposed SIF metric calculation that is more leading than lagging is the SIF potential rate. This metric identifies high-energy events that could have led to SIFs. It is calculated as follows: (SIFp Events × 200,000) ÷ Hours Worked.

16.8 SIF CONSENSUS STANDARDS

Additionally, the National Safety Council (NSC), in coordination and cooperation with the American Society of Safety Professionals (ASSP), has also been developing and updating a consensus standard for the sake of standardization of definitions, measurements, mitigations, and approaches to dealing with SIFs. So, between NSC, ASSP, CSRA, IEC, and ISO, the following are the consensus standards that have been leveraged for SIF intervention (this list is not exhaustive):

- ANSI/ASSP Z16.1–2022 – Safety and Health Metrics and Performance Measures.
- ASTM E2920 – Guide for Recording Occupational Injuries and Illnesses.
- ASTM E2921 – SIF Rate calculation, which focuses on life-altering outcomes rather than minor injuries. SIF Rate = (SIF Events × 200,000) ÷ Hours Worked.
- IEC 61511/ISA 84.01 – Foundational for the design and lifecycle management of safety-instrumented systems. They define how SIFs fit into safety lifecycle approaches, encompassing framework, hardware/software requirements, and verification of safety integrity levels.
- ISO 13849 – Focuses on functional safety in machinery, detailing safety-related control parts and validation, but applies more to machinery than process industries.
- ASTM WK92349 – New Guide for SIF Risk Assessment (NSC/ASSP collaboration mentioned earlier). I have the privilege of being a part of this endeavor.

I think it is extremely important to better understand the key components and main contributors to serious injuries and fatalities (SIFs), because this understanding will help prevent them and build controls that fail safely. This two-pronged approach, fueled by the newfound focus on SIFs, will help bring more workers home safe in the United States and around the world. Another reason why I believe that the focus on SIFs is critical today is because it is another step toward broadening our perspective on safety. Focusing on SIFs helps our organizations to better appreciate the multifaceted nature of holistic and integrated safety.

16.9 LEVERAGING TECHNOLOGY (AI+) IN SIF PREVENTION AND RECOVERY

With rapid technological advancement, especially in the areas of artificial intelligence (AI) and its various applications, such as machine learning (ML), computer vision (CV), knowledge-based systems, and natural language processing (NLP), these technologies can and should be leveraged in our struggle with SIFs. When combined with multiple wearable applications, the Internet of Things, exoskeletons, and the like, we have a formidable arsenal to use in this endeavor. For example, by employing AI technologies such as computer vision and knowledge-based systems, it is possible to continually evaluate workplace environments through sensors, cameras, and wearable technologies. This constant surveillance aids in providing timely warnings to workers when they are at risk of significant hazards that can result in SIFs. AI can support the collection and analysis of workers' data on their physical situation, mental stability, and emotional well-being, thereby facilitating enhanced monitoring and performance improvement. Using this data, AI systems can provide early warnings and interventions to help prevent burnout, depression, anxiety, or other conditions with SIF potential. Additionally, you can train the AI to recognize SIF potential events and hazards and then utilize it to both process all your historic incident data in order to identify trends you may have missed and offer you predictive capacity for the types of hazards, near-misses, and incident potential situations that can lead to SIFs. This gives you the ability to forecast potential hazards before they can become imminent threats.

16.10 COLLABORATION – KEY TO DEFEATING SIFS

At the end of the day, successfully defeating rising SIF rates is going to take a collaborative effort – all of us working together. What do I mean? I mean every stakeholder within the organization, from operations to safety to procurement. And I mean different organizations finding ways to cooperate and share.

I realize this gets sensitive really quickly because no organization likes to share its dirty laundry or show its biggest pain points, especially when such a revelation can lead to further regulatory scrutiny, litigious activity, or contractual consequences. But that is what it is going to take to turn the tide of SIFs.

To this end, I'm so glad to see the oil and gas industry leading the way in the effort to reduce SIFs by holding Safety Shares. As an example of this, the American Petroleum Institute is hosting a quarterly Onshore Safety Alliance gathering of operators and contractors throughout the United States, where they can talk about these significant events and learn from one another. This is a great opportunity for safety professionals to come together and discuss the incidents, trends, learnings, and mitigations. CSRA, mentioned earlier in this chapter, has been holding similar conversations in the construction industry, which is also great. But more cooperation is absolutely needed in this arena. If we don't, people will keep dying.

BIBLIOGRAPHY

Dekker, Sidney. (2025). Safety Theater: How Success Can Mask Growing Safety Risks. *PSJ*, October, 28–34.

Goncharenko, Simon. (2024). *Save Lives: Pushing Boundaries in Human Factors*. Kindle Press.

Goncharenko, Simon. (2025). *Operationalizing Twenty First Century Safety: A Humancentric Practical Guide*. CRC Press.

Hallowell, Matthew, Quashne, Mike, Salas, Rico, MacLean, Brad, & Quinn, Ellen. (2021). The Statistical Invalidity of TRIR as a Measure of Safety Performance. *PSJ*, April, 28–34.

Kotek, Lubos, Nosek, Ales, & Bartos, Vojtech. (2018). Safety Metrics of Performance for Small and Medium-Sized Enterprises – Case Study. *MM Science Journal*, (01), 2333–2337.

Trivedi, P., & Alqahtani, F. M. (2024). The Advancement of Artificial Intelligence (AI) in Occupational Health and Safety (OHS) across High-Risk Industries. *Journal of Infrastructure, Policy and Development*, 8(10), 6889.

17 Corrective Actions as the Focus of Incident Investigations

Chris Gafford

17.1 INTRODUCTION

"The injured employee didn't follow company policy. Corrective action is to reiterate the importance of following policy with all employees." This is a quote. It comes from a contractor's Incident Investigation and Corrective Actions report after they narrowly escaped killing two guys on one of my company's locations. Anybody can see that it leaves a lot to be desired.

17.2 REAL-LIFE INCIDENT EXAMPLE

Fortunately for everybody involved, the injury was a simple sprained ankle instead of two grieving families. During a drilling rig move, one of the pole truck drivers was trying to be helpful in the crowded location by backing his truck up into the pasture. He knew that the poles were still in the extended position. He knew that they had discussed the powerlines at the edge of the location during the morning's prejob tailgate safety meeting and even noted the powerlines on the job safety analysis (JSA) sheet. Most importantly, he knew that his company policy stated that no pole trucks leave the location with the poles extended.

Ironically, this driver was the reason for the policy in the first place. Approximately 8 years earlier, he had driven down a county road with his poles extended and torn down a powerline. In that situation, the lines broke, the fuses blew, and the only negative consequence was a small bill for line repair and a badly bruised ego. But it had all the ingredients for a fatal incident.

Now, he had done it again. Granted, the circumstances, the date, and the location were different, but the principle was the same – a powerline was engaged with the poles of his truck. Just like before, the lines came down. Unlike before, the fuses did not blow, and now this driver and his helper (swamper) were sitting inside a truck with live power lines lying across the top.

The swamper, a recent high school graduate, was working his first "real" job. He was an athlete in high school and aspired to play football in college. But when his collegiate aspirations didn't pan out, he did what many young men in the Permian Basin do – got a job in the oilfield. Before he had learned all his coworkers' names or even gotten rid of the dark green Short Service Employee hardhat, he was sitting in a

DOI: 10.1201/9781003679448-17

truck draped with a live power line. Lacking the training, experience, and cool-headedness of his partner in the truck, the young man acted in fear and immediately jumped out of the truck when he saw the sparks of the lines breaking. His athletic ability and survivor instinct, now powered by fear for his life, propelled him far enough from the truck that he somehow made it out without being electrocuted. He was uninjured. During the ensuing interview, the young man would admit that he didn't even know it was possible to contact the truck and become the grounding point that the electricity was seeking, which begs the question of the training he received or didn't receive. But that's another point for another chapter.

Meanwhile, the truck driver has been in this situation before. At 60+ years of age, he had a little more life experience. That's especially true when you consider that 30+ of his work years had been dedicated to the oil and gas industry. So, the driver calmly waited in the energized truck as electricians made their way to the area to disconnect the lines.

But then the tires of the truck began to smoke. Seeing the smoke from the tires and factoring in the kid's success in leaving the truck, the driver now decided it was time to make his exit as well. Against the advice of all those around him witnessing this harrowing event, the truck driver jumped as far as his aged knees and bad back could muster. The phrase "He hit the ground running" was true of his swamper but not exactly accurate for this gentleman. The driver hit the ground rolling. His right boot landed on a rock and immediately rolled into a painful sprain, as his knees buckled, and he hit the dirt like a sack of deer corn.

Seeing a grown man fearing for his life and trying to run on old knees and a newly sprained ankle can make for a hilarious YouTube video. However, lives were on the line, so this was no laughing matter. In the end, both men survived, unlike several previous incidents in the Permian Basin, where equipment struck live powerlines, resulting in the loss of life of the equipment operators.

17.3 TO LEARN OR NOT TO LEARN FROM THE INCIDENT – OUR CHOICE

When I got the call about the incident, after breathing the same sigh of relief that everybody on location breathed seeing that the two men survived, I asked the part-time safety representative for the trucking company to conduct a root cause investigation and provide corrective actions to prevent a recurrence. This is simply a standard procedure with all injuries or significant near-misses in the industry. I was aware that she was a new safety professional. She worked part-time for the trucking company, while her other concurrent responsibility was attending a local college in pursuit of an official safety degree. Of course, besides the degree, the idea was that these college courses were also adding knowledge to the student about what it means to be a safety professional.

I also knew that the trucking company was a small and unsophisticated contractor. This is the kind of company where the owner drives one of the trucks, and if you call the number on the door of the truck, the guy driving it might be the one that answers. With all that in mind, I still held out faith that the safety representative had studied hard in her classes and would be able to provide a thorough and well-researched investigation with meaningful corrective actions. I was wrong.

This scenario of contractors not having the ability, the desire, or both to do a thorough investigation or learning event and identify corrective actions and controls, aimed at informing prevention abilities and ensuring recovery from future unwanted events, has played out over and over during my career. It's difficult to put a finger on exactly where the problem lies. Sometimes it's ignorance. Sometimes it's apathy. Sometimes it's inability. Sometimes it's . . . , and sometimes it's a combination of all of the above.

17.4 IDENTIFYING CAUSAL FACTORS TO PREVENT RECURRENCE AND BOLSTER CAPACITY TO RECOVER

On the seemingly rare occasion that the contractors do turn in an actual investigation report, many in the onshore oil and gas industry seem to opt for the "Five Whys" Root Cause Analysis (RCA) method. In my experience, this method is highly ineffective and rarely leads to the true root cause of the incident. You might call it the uninformed person's root cause analysis.

Of course, I am aware of the ongoing challenge to the whole concept of the root cause in favor of the more nuanced approach that considers the complex interplay of contributory factors. For more information on this debate, see Hummerdai, Latino, and Pollock resources listed in the Bibliography section, as well as Gantt's Proactive Learning chapter in this book. They should be able to initiate your research in this area if you're curious.

Because the focus of this chapter is on the importance of corrective actions and not on either proving or disproving the legitimacy of the root cause concept, I will continue here with the traditional assumption that still pervades the onshore oil and gas industry that there is such a thing as root cause, that it is identifiable, and that discovering it can contribute to the greater possibility of future incident prevention, even if not a full elimination.

The basic idea of the Five Whys method is for the investigator to keep asking "why" past the human error answer and the other easy responses, in order to get to the systemic issues that may be at the root of what caused this unwanted event. The hope behind this exercise is that by mitigating this root cause, future recurrence can be prevented. Some are beginning to question whether the logic behind this exercise may be too linear and overly simplistic, given the complex and multifaceted modern workplaces, as stated earlier. Be that as it may, given that this is still the prevalent method of post-incident response and learnings across a number of different industries, including oil and gas, construction, and facility management, to name a few, we will flesh it out further here, while encouraging the reader to do some additional research on the topic.

Even though the concept of learning from our unwanted events is so simple, what I have found over the past three decades of experience in the energy industry, the implementation of it is often less than desirable. Many, if not most, of our incident learnings lack real substance. We don't have the taste for asking "Why" enough times to get to the systemic issues, and, as such, no sufficient corrective actions are produced, no real mitigation of future occurrences takes place, and no erection of additional controls is done. In my practical mind, using the "Five Whys" method is a lot like putting a Chevy Corvette logo on a Ford Pinto. You can call it fancy and maybe trick a few people, but the performance just isn't there. Safety professionals must do better.

Part of the responsibility of the safety professional is to educate the organization in the best ways to learn from incidents, as well as to do everything in their power to ensure that causal factors are identified, mitigated, and proper controls are erected to prevent recurrence and bolster the capacity to recover. The latter part is usually defined as the corrective actions. Failing to do so exposes the frontline workers to hazards. Even worse, it's leaving them exposed to known hazards, which is, of course, illegal and unethical.

One of the problems is that contractors often lean on inexperienced safety professionals simply due to their title. They hire a "safety guy" without much experience and assume everything will be alright now that they have a safety department. It's a bad choice, however, to leave the safety guy on an island without support, especially when it comes to incident investigations. It should be a group activity, not a solo exercise.

The safety professional is rarely a subject-matter expert in the company's operations. In situations where the safety rep does have an intimate knowledge of operations, it's usually because the company took a good operations guy and put him in a safety role. That transfer rarely comes with the proper education or training. The unfortunate result of this lack of understanding caused by the improper training in either case is a safety professional who doesn't know enough about operations or an operations person who doesn't know enough about safety. Now, if you put the two together, suddenly, you have the beginnings of a solid investigation team with the knowledge to cover the gaps.

Another challenge that contractors regularly face when doing an incident investigation is the hard questions that will have to be answered. While investigations aren't meant to point fingers or lay blame, that's often what happens, either directly or indirectly. This puts contractors in a tough position when they are forced to admit to their customers that the real reason for the incident was the hiring client's own shortcomings.

17.5 CORRECTIVE ACTIONS – WHAT YOU DO AFTER LEARNING FROM AN INCIDENT

Once the incident investigation is over, you're faced with the hurdle of coming up with corrective actions that will actually make a difference. This is where unskilled or inexperienced safety professionals will come up with useless corrective actions like "retrain crews on the procedure" or "discuss the incident in safety meetings." You might even see something as simple as "send out safety alert." The question is – are any of those actual corrective actions? Admittedly, it's important to share the details of the incident and results of the investigation with the rest of the company. Unfortunately, however, sharing the information falls far short of the ultimate goal, which is to prevent it from happening again or identify appropriate controls that will keep a similar event from hurting or killing our employees.

Some contractors can continue to have a seemingly successful business even if they never do a thorough incident investigation or develop quality corrective actions. They continue to thrive due to several different factors like lower pricing, personal relationships, or lack of competition. The sad reality is that while the company may survive, its employees are still in danger. A strong balance sheet or busy schedule does not always equal a safe workplace.

There are some things that contractors can do to provide their customers, the hiring clients, with solid investigations that will reaffirm their faith in the contractor's commitment to keeping their employees safe.

1. Make sure your safety professional or whoever will be doing the investigation has the proper knowledge and training. Knowing how to ask the right questions and what questions are right does not come naturally, so one must be taught. Additionally, there are several solid root cause methods available, and training can be arranged on any of them. Oh, and when you send your safety professional to meet with the client, make sure this individual is armed with the knowledge and the ability to represent you well. After all, the safety lead is, at this point, the face of the contractor when explaining how an incident happened.
2. Incident investigation is rarely beneficial with an individual approach. Even detectives at law enforcement agencies work as a team to investigate serious crimes. Include multiple people with differing skill sets in your investigation team. A foreman, an executive, and even a frontline worker all possess knowledge and experience that can help in the investigation.
3. Approach incident investigation as an opportunity to improve your company and the lives of your employees. The purpose of finding the root cause is to prevent it from happening again and to erect controls that will make your operations safer in the future. There is a saying that OSHA regulations and safety policies are written in blood. That doesn't have to be the case. If your organization takes a proactive approach to investigating not just accidents but near-misses as well, there's a good chance you can prevent spilling the blood that would normally write the next policy or procedure.

The important actions after an incident don't stop once the investigation is complete. That's just the first step. Once you know what happened and why it happened, then you find a way to never allow a similar event to result in an injury or death.

Imagine being the truck driver at the beginning of the chapter who repeated the same mistake from 8 years ago. Not only will that put a large dent in his pride, but it could have cost him his life. Now imagine being in management or possibly an ownership role of a company and not instituting the right corrective actions or controls after the first time an incident took place. The incident is repeated, but this time with deadly consequences. Nobody wants to be in a position where they have to tell a family, coworkers, OSHA investigators, or attorneys in a courtroom that they were aware of the hazard but failed to find a permanent and meaningful mitigation.

The sad truth is that the corrective actions developed by inexperienced safety personnel or without the help of subject-matter experts regularly fall short of protecting workers. This is especially true when meaningful corrective actions aren't readily apparent or the people responsible for developing them don't have the experience or expertise needed. Remember the quote from the incident report mentioned earlier? "Corrective action is to reiterate the importance of following policy with all employees." While this may be a good idea and certainly a step that needs to be taken, it's highly unlikely that simply "reiterating" anything will change behaviors. This simple approach certainly isn't going to shift the culture of the organization to safer operations.

Just like with the incident investigations, there are several things to keep in mind to develop true and meaningful corrective actions after an incident or near miss.

1. The whole purpose of a corrective action is to prevent the recurrence of unwanted events or their ability to hurt or kill. It should be approached as such. Too often, safety professionals are aiming to put something down that will satisfy the hiring client, as opposed to focusing on what really matters, which is making sure their employees are safer after the incident than they were before.
2. Any corrective actions coming out of an incident investigation should be considered with as much thought and weight as the investigation itself. To use the football analogy, don't get within the eyesight of the end zone just to drop the ball too early. Finish out the process, and everyone involved, including the hiring client, will be better off for it.
3. Tracking corrective actions to completion is imperative. Not doing so is a trap we all fall into. Robert Burns' poem "To a Mouse" teaches us that "The best laid plans of mice and men often go awry." Ingenious and iron-clad corrective actions can be developed, discussed, and planned, but if not tracked to completion, they will inevitably fall through the cracks. Assign the corrective action to a person or department, assign a completion date, and assign a follow-up to review the implemented actions and their effectiveness.
4. Spreading details of the incident internally should be the bare minimum. It's important to be an information-sharing organization so that all employees ultimately know what happened, why it happened, and how to keep it from happening again or how to keep from allowing it to hurt us the next time. As mentioned before, "reiterating" some old policy to people who are already aware of it just doesn't do much good. It certainly doesn't move the needle on safety culture.
5. Many times, training is listed as part of the corrective actions. That's great, and it's usually warranted. The important thing about training is that it must be actual training. You can't just mention an incident during the next safety meeting and consider your employees trained. Training should be treated as an event. There should be a set time, place, and curriculum that covers all aspects of whatever the topic is. Due to time constraints and scheduling conflicts, contractor companies rarely have the opportunity to deliver real training. It regularly ends up being more akin to a slide or two in the

monthly safety meeting PowerPoint. Everybody understands the difficulties of gathering up employees, scheduling conflicts, etc. It's not easy. However, if you consider that an incident may be an alternative to skipping out on needed training, suddenly, training is no longer a huge burden.

6. When developing corrective actions, one of the first considerations should be the "hierarchy of controls." Go down the list and use the highest hazard control that can actually be implemented.
   a. Elimination
   b. Substitution
   c. Engineering controls
   d. Administrative controls
   e. PPE

For whatever reason, organizations seem to get these controls backwards. Maybe it's due to ease of implementation or just simplicity. It's important to keep in mind that adding PPE or putting in new administrative controls are at the bottom of the hierarchy and are somewhat useless compared to the controls above them.

You may have heard during a motivational lecture or an inspiring half-time speech that "Failure is not an option." That's incorrect. Failure is always an option.

As a matter of fact, failure is usually the easiest and most readily available option. For that reason, great concern, care, and attention to detail must be used in the investigation process as well as the corrective action portion in the aftermath of any incident. Failure to do the job well can, and probably will, result in a repeat of the incident. Likewise, failure to do the job well can, and probably will, put workers in unnecessary danger. If there is one place where employers should try to remove failure as an option, this is it. Give the investigative process the amount of seriousness and professionalism needed to avoid failing your company, your customers, your employees, and their families.

BIBLIOGRAPHY

Hummerdai, Daniel. (2013). *The Evil of Roots*. https://safetydifferently.com/the-evil-of-roots/

Latino, Bob. (2022). *Do Learning Teams Make RCA Obsolete?* https://reliability.com/resources/articles/do-learning-teams-make-rca-obsolete/

Pollock, Richard. *Human Error: There Is No Root Cause*. https://pdfs.semanticscholar.org/4944/d033bad78347f0cb15d260eb8d446b8cabc3.pdf

18 Leveraging AI and Other Technology for Smarter Contractor Management

Simon Goncharenko

18.1 TECHNOLOGY TODAY

Our world is shaped by technology. Most of us cannot imagine our lives without the ease and assistance of technological tools, including artificial intelligence (AI), Internet of Things (IoT), wearable tech, and a multitude of app-based aids. Once a futuristic concept, many of these tools have moved to mainstream technology, reshaping industries and impacting daily life globally. The following list contains just a small statistical sample of how these technological advances have been adopted worldwide:

- AI has been progressively incorporated into various applications through machine learning (ML), computer vision (CV), knowledge-based systems, and natural language processing (NLP).
- General AI usage by organizations around the world jumped from 55% in 2023 to 78% in 2024.
- Generative AI usage in at least one business function by the reporting companies nearly doubled in the same time period, reaching 71% in 2024.
- Many of these companies are expanding their technological usage to at least three different functions within their system.
- IT, marketing and sales, and service operations have the highest rates of AI adoption.
- In 2024, corporate AI investment reached $252.3 bn globally.
- Occupational exoskeletons are designed to minimize the risk of back and shoulder injuries in workplaces where traditional ergonomic solutions fall short.

18.2 AI IN CONTRACTOR MANAGEMENT

The list of statistics can go on and on, but the moral of the story is clear – technology must be utilized for businesses today to stay relevant, competitive, and profitable. This principle certainly applies to contractor management. The list of options and tools in this space is rapidly growing and expanding. The general principle here is that both hiring clients and contractors can choose from the existing range of apps and tech or create their own internal solutions that align more closely with their intended purposes and processes.

DOI: 10.1201/9781003679448-18

Let's double-click on AI for a moment, utilizing some of the research and findings provided by Trivedi P and Alqahtani FM (see Bibliography). Artificial intelligence encompasses the capabilities of computer systems or robots to execute tasks like language comprehension, image identification, decision-making, and problem-solving, with conventionally mandating human intellect. AI-enabled devices, like robots or drones, are increasingly used for inspection, maintenance, or repair operations in perilous settings such as mines, power plants, telecom towers, refineries, and construction sites, enhancing safety and efficiency.

But before discussing individual tools, perhaps a more general question of how AI can be incorporated into contractor management should be answered first. AI tools and technologies can be integrated into streamlining contractor vetting, managing contractor training, and monitoring contractor compliance as a starting point. Drilling down further, AI tools could be leveraged to identify and analyze patterns in contractor performance, such as common safety violations, injury precursors, etc. Additionally, AI could offer predictive analytics to flag potential compliance issues and risks. It can also proactively manage risks and track contractor accountability.

Of course, as you map out incorporation of technology tools, you will need to plan on creatively overcoming common challenges as well. Your organization may be resistant to change. To overcome this, you will need to schedule training sessions on the value of AI. Then there is a challenge of ensuring system compatibility between your current tools and the new AI technology. To overcome this, you will want to choose scalable platforms with strong integration capabilities, like ERM, ChatGPT Enterprise, GROK, You.com, etc. Equally important – and sometimes the most critical – challenge introduced by technology is data security. To overcome this challenge, you will want to select trusted vendors who offer advanced cybersecurity measures and protocols.

18.3 USE-CASES OF AI IN CONTRACTOR MANAGEMENT

If you are still unclear about just how these AI and tech tools can help you, here are some specific examples of contractors adopting these solutions, as outlined in Trivedi P and Alqahtani FM's 2024 article, "The Advancement of Artificial Intelligence" (see Bibliography):

- AI-enabled devices, like robots or drones, can be used for inspection, maintenance, or repair operations in perilous settings such as mines, power plants, telecom towers, refineries, and construction applications.
- AI can support extensive collection and analysis of workers' data concerning their physical situation, mental stability, and emotional well-being, thereby facilitating enhanced monitoring and performance improvement. With this data, AI systems can provide early warning and interventions to help prevent conditions like burnout, depression, or anxiety.
- AI can play a pivotal role in occupational health and safety (OHS) training by crafting realistic, immersive simulation scenarios, particularly for emergency responders. Such simulations allow workers to better visualize and effectively manage various emergency situations at work.

- AI can facilitate the automation of dangerous or physically demanding tasks using robots or collaborative robots (cobots), which can significantly reduce the exposure of workers to hazardous conditions.
- AI can be leveraged to analyze vast quantities of occupational data, which facilitates the identification of patterns, trends, and potential OHS risks. This gives you the ability to forecast potential hazards before they can become imminent threats. This predictive capability not only helps in preempting accidents and injuries but also supports proactive measures in managing workplace safety.
- Occupational exoskeletons, mentioned earlier, can support workers and have received positive early-use feedback. Workers wearing these devices reported less discomfort, fewer injuries, and reduced workers' compensation costs. Studies have shown that occupational exoskeletons can lessen muscle strain and fatigue during physical activities across various sectors, including logistics, construction, manufacturing, military, and healthcare.
- Another potential CV application can analyze real-time video footage from cameras and drones on installation sites where contractors are working. The system continuously verifies that contractor employees are using proper PPE and following established safety protocols. Noncompliant behaviors are flagged immediately, enabling rapid corrective actions, which in turn ensure proactive incident prevention and culture improvement.
- Additionally, CV can also be deployed via the incorporation of thermal cameras that detect heat stress among workers. This allows continuous monitoring of body temperatures, enabling timely interventions like cooling breaks or the provision of suitable PPE.
- Then there is an option to leverage predictive scores by integrating AI algorithms into your contractor management system, which aggregates data from multiple sources, including training records, certifications, incident logs, and performance metrics. The AI system can then evaluate historical and real-time data to generate risk scores for each contractor, highlighting potential areas of noncompliance. High-risk contractors can be scheduled for additional training or audits, minimizing the likelihood of safety incidents and improving operational efficiency.

As with all other tools, AI incorporation is not without its challenges, not the least of which is the high cost of implementation, lack of skilled workforce trained in these technologies, and social concerns related to data privacy, worker surveillance, and algorithmic bias. These issues not only complicate the incorporation of technology, but they may also introduce new risks, such as cyber threats and psychosocial stress, associated with human–machine interactions. To overcome these challenges, your organization must exercise an interdisciplinary approach that includes your safety, human resources, legal, operational, financial, and other relevant stakeholders.

18.4 AI AND TECHNOLOGY TOOLS FOR ADMINISTERING INDIVIDUAL ASPECTS OF CONTRACTOR MANAGEMENT

While I will attempt to provide a short summary of the types of technology tools that are available today, I am certain that even in the time that will pass between the writing of this chapter and its publication, my list will be dated, and more research will be needed to bring it up to date. For that reason, instead of providing you with the names of the apps, I will instead focus on giving you ideas for the types of services that you can procure, leaving the final step of locating the right tool up to you. As I mentioned in the introductory chapter of this book, there is a technological tool in existence today that can help you manage every individual aspect of contractor management. For example:

- If you need help with creating localized contracts in different global geographies, managing payments in multiple currencies, and ensuring compliance with in-country regulations, deel.com is an example of a tool that can help you.
- If you need assistance with hiring specifically trained, educated, or certified personnel in various global locations, remote.com is an example of a tool that can help you in this endeavor.
- If you need help with onboarding your construction industry contractors, lining up certain training requirements for them, and such, gocontractor.com is an example of a tool that can help in this space.
- If you need help with auditing and conducting assessments of your contractors, ERM is an example of a tool that you can use.
- If you need assistance with aligning with your contractors on their investigation of incidents, inputting corrective actions, and managing the closure of those findings, KPA is an example of a tool that can help.
- If you need assistance with managing drug and alcohol policy creation and adherence, you can check out DISA.com.

18.5 END-TO-END CONTRACTOR MANAGEMENT SUITE

At the same time, there are also global providers who focus on delivering the complete suite of contractor management tools under one roof, including selection, prequalification, insurance, onboarding, training, monitoring, and auditing your contractors. Perhaps the best example of this type of solution is Veriforce, which is the largest global provider of contractor management solutions, currently helping over 3,000 hiring clients to manage over 140,000 contractors in 140 countries around the world.

At the end of the day, the technology is here to stay, and smarter clients and contractors embrace it, leverage it, and incorporate it into their daily protocols in order to stay competitive and profitable, while protecting the health, safety, and well-being of their personnel.

BIBLIOGRAPHY

Trivedi, P., & Alqahtani, F. M. (2024). The Advancement of Artificial Intelligence (AI) in Occupational Health and Safety (OHS) across High-Risk Industries. *Journal of Infrastructure, Policy and Development*, 8(10), 6889.

19 Qualification of Contractor Pipeline Personnel

Thomas N. Meek, Jr.

19.1 INTRODUCTION TO PIPELINE REGULATIONS

In the United States, pipelines are a safe and reliable form of transportation and, as such, are regulated by the US Department of Transportation (DOT), administered through the Pipeline and Hazardous Materials Safety Administration (PHMSA). PHMSA has the responsibility for developing guidance and enforcing regulations for the safe operation of the country's pipeline infrastructure. The purpose of these regulations is to reduce the risk of serious incidents that can harm people, property, and the environment.

Through the years, agencies such as the National Transportation Safety Board (NTSB), which is an independent agency that investigates civil transportation accidents, discovered many of the incidents were directly linked to human error. In a 1987 NTSB letter to the Research and Special Programs Administration (RSPA), the NTSB pointed out that human errors that could be linked to inadequate training had caused or contributed to the severity of many pipeline accidents.

In August 1994, the RSPA proposed qualification standards for pipeline personnel "who perform, or directly supervise those persons performing, regulated operation, maintenance, and emergency-response functions." The rulemaking notice referred to multiple instances of recommendations from the NTSB that the agency requires operators of pipelines to develop and conduct selection, training, and testing programs to annually qualify employees for all activities necessary to comply with pipeline safety standards. Some, but not all, of these recommendations are in today's regulations.

Key moments in the regulatory journey include the first statute regulating pipeline safety, the Natural Gas Pipeline Act of 1968. In 1979, Congress included standards for the safe operations of pipelines for the transportation of hazardous liquids. The Pipeline Safety Act of 1992 was to increase safety for humans and the environment. It also mandates operator qualification (OQ) programs. Operators were required to have their written programs by April 27, 2001, and complete the qualification of workers performing covered task(s) by October 28, 2002.

DOI: 10.1201/9781003679448-19

19.2 TRAINING AND QUALIFICATION

Before the OQ Rule became effective, there were many steps in between the Notice of Proposed Rule Making (NPRM) and the process of negotiating what the final rule would include. One point of contention at the time was the requirement for training. In those early days, the industry had many tenured, experienced employees who were already considered qualified because they were completely capable of performing their daily jobs.

The time and expense of training qualified employees didn't make sense at that point. The consensus was to withdraw a strict training rule and focus on qualification verification instead. The logic was "qualification" is an end, while training is merely a means to an end. In turn, "qualification" results from an evaluation, while training does not necessarily always result in a qualification. So, today we have a process that verifies qualification via evaluation without a strictly prescribed mandate for training. A worker can qualify without any formal documented training and go to work.

For this reason, in the pipeline, unlike many industries, training does not equal qualification. An analogy often used to explain why training doesn't equal qualification is the common experience of earning a driver's license. A person can complete all the required training, but until they take the practical exam and successfully demonstrate their knowledge, skill, and ability to drive a vehicle, they are not qualified. Next, we will examine when training is required and how to identify work classified as covered tasks.

19.3 COVERED TASKS

The Code of Federal Regulations CFR 49 Transportation § 192.805 (h) & §195.505 (h) reads as follows: "to provide training, as appropriate, to ensure that individuals performing covered tasks have the necessary knowledge and skills to perform the tasks in a manner that ensures the safe operation of pipeline facilities."

Except for a few states with training requirements, "providing training when it's appropriate" has been defined by each operator. Examples of when training is appropriate for most operators are prior to an initial qualification attempt, after a failed qualification attempt, prior to a subsequent qualification attempt, in the event of an incident, or if the individual hasn't performed the task for some time.

Having spent the past three decades working in a variety of different aspects of the pipeline industry, I have the experience and qualifications necessary to have formed my own opinion on the current state of training. In my view, the industry assumption from 20 years ago is no longer viable. While two decades ago there may have been a surplus of tenured workers able to perform their daily tasks without training because they were already considered qualified, such is no longer the picture with the generational changes. The time and expense of training new-to-industry employees does make sense today.

If human errors can be linked to inadequate training that causes or contributes to the severity of many pipeline accidents, then training is "appropriate" for first-time gas and oil industry employees to acquire the prerequisite knowledge, skills, and abilities needed to competently perform their jobs and return home safely from highly hazardous work. The industry has two factors that make training essential. The retirement of skilled workers with institutional knowledge, combined with the increasing demand for energy infrastructure, is creating new jobs.

Currently, the baby boomer generation is reaching its retirement age. As these legacy workers retire, they take with them their knowledge, skills, and abilities, which are critical for safely building and maintaining pipelines. This trend has made it difficult to backfill these positions with properly trained and qualified new hires at the same rate as retirements. There is a growing gap because of workforce trends and evolving regulatory requirements, which increase the challenges pipelines face as they continue their mission of safely and reliably delivering energy.

The growing footprint of new pipeline projects will increase the need for skilled gas and oil workers. Pipeline growth will be necessary to meet the demand for power generation, liquified natural gas (LNG) exports, and facilities supporting modern technologies such as data centers. At the heart of today's challenge for pipeline operators is maintaining a qualified workforce that can provide safe uninterruptable service to their customers. The gas and oil industry faces a gap in its ability to hire, train, and qualify workers who are new to the industry. The hiring challenge is the same for contractors and the pipeline operators they support.

Hiring, training, and qualifying workers isn't optional. Pipeline operators have many challenges to maintain operational integrity and regulatory compliance. With simultaneous factors like the retirement of the workforce, evolving regulations, technology, and the demand for safe, clean energy creating more gas and oil jobs, companies can prepare by taking proactive steps to solidify their OQ programs.

We can look at the Code of Federal Regulations to find the minimum requirements for qualifying pipeline personnel. The regulations serve as a foundation for an operator to build and mature their OQ program for the future. While no regulation can completely eliminate human error and accidents, and other chapters comment on these areas in more detail, the CFR 49 Transportation § 192.805 (h) & §195.505 (h) rule was put in place to reduce risk and impact of incidents by adding an additional layer of safety. This additional layer requires pipeline operators to develop and implement a program for initial qualification and ongoing subsequent refresher qualification to ensure there is no lapse in an individual's ability to competently perform the job and recognize and react to anything that may go wrong while performing their work.

The intent behind the OQ rule is to ensure that the workforce on regulated pipelines is qualified. Qualified workforce, according to this line of reasoning, will reduce the probability and severity of pipeline incidents or accidents caused by human error. Therefore, qualifying is not intended to be a one-time event, but a process that continues throughout an individual's career.

Identifying these tasks isn't as intuitive as one might think. Many life-critical tasks or jobs occur every day on the pipelines that aren't classified as OQ tasks. For example, working in a confined space or from heights can put one's life in danger. Another non-OQ task that is one of the most hazardous things we do in the industry is driving a vehicle. So, what is an OQ task? Let's look at the regulations for guidance.

19.4 OQ TASK

Within the written OQ program, an operator must identify the jurisdictional tasks performed on their pipeline facilities and set ongoing subsequent requalification intervals to ensure that there is not lapse in qualification status. Operators must follow a prescribed method to set minimum requirements, commonly called the four-part test, to determine their operation's specific task list. In CFR §192.801 & §195.501, "Qualification of Pipeline Personnel," the criteria are provided for operators to make this determination. It is an activity that:

1. is performed on a pipeline facility;
2. is an operations or maintenance task;
3. is performed as a requirement of this part; and
4. affects the operation or integrity of the pipeline.

For clarification, "pipeline facility" includes the pipelines themselves. And "performed as a requirement of this part" points to anything specifically called out in the regulations. These are the federal minimum requirements to determine a task list. It is noteworthy to point out that some industry standards and state requirements have simplified the determination of a task list by focusing on safety and integrity.

The reason every operator does not have the same "off-the-shelf" task list is elucidated by the four-part test. For example, each operator has unique facilities, geographical considerations, population density, and types of equipment. The main reason each operator can have a different task list than their peers is due to their differences in the operations and maintenance (O&M) manuals. If all O&Ms were the same, it would be reasonable to expect that operators' task lists would also be similar. The truth of the matter is that operators have unique operations policies and procedures, making their O&Ms unique and, therefore, their task lists one of a kind.

This has been an ongoing source of contention as operators are incorrectly expected to have the same task list as their neighboring pipeline peers. Simply put, if the operator's O&M describes the work, then they should have a corresponding task(s). And if the operator is performing the work/task on a jurisdictional pipeline, then it should be part of their O&M. You will often hear the term "covered tasks," which means a task meeting the four-part test criteria. Now that we have described the activity defined as a task, let's explore how an individual becomes qualified.

19.5 QUALIFIED WORKER

For a jurisdictional pipeline worker to be qualified, they must: (1) have been successfully evaluated; (2) be able to perform the assigned or identified covered task; and (3) recognize and react to abnormal operating conditions. With regard to an evaluation, it is defined as a process identified by the operator that must be documented to determine an individual's ability to perform a covered task by any of the following:

1. Written exam,
2. Oral exam,
3. Work performance history review,
4. Observation during:
   a. Performance on the job
   b. On-the-job training (OJT), or
   c. Simulation.
5. Other forms of assessment.

It is important to note that work performance history and on-the-job performance may not be used as the sole means of evaluation. PHMSA also provides additional guidance in the form of frequently asked questions (FAQs). Within the FAQs, there is a phrase about the intent of qualification, where an individual demonstrates knowledge, skills, and abilities as part of the act of qualifying. Taken together, an individual should receive appropriate training and be evaluated to confirm their knowledge, skills, and abilities.

The step mentioned earlier ensures that the worker can perform the assigned or identified covered tasks. The third requirement for a qualified worker is to be able to identify and respond appropriately to abnormal operating conditions associated with the task they are performing. But what is abnormal? According to the CFR §192.801, an abnormal operating condition means a condition identified by the operator that may indicate a malfunction of a component or deviation from normal operations that may: (1) indicate a condition exceeding design limits; or (2) result in a hazard to persons, property, or the environment.

The OQ requirements are not optional. To perform work on a regulated pipeline, an individual must be qualified prior to performing the work. It is the operator's obligation to ensure all individuals are qualified in accordance with the rule. This is true regardless of whether you are an internal pipeline operations employee, a contractor employee, or a subcontractor performing a covered task on the pipeline. Nonqualified workers are only allowed to perform covered tasks while being directed and observed by someone who is qualified and can intervene in the event of abnormal operating conditions. The term for overseeing nonqualified individuals is referred to as span of control (SOC). The SOC allowance wasn't intended to be a workforce multiplier to put nonqualified people on the job. It was a logical method to allow for OJT that could lead to qualification on a regulatory pipeline under the supervision of a qualified individual.

Nobody is exempt from being qualified prior to independently performing a covered task. There is no grace period for pipeline operators to confirm qualification, regardless of size of operation. Each pipeline operator has the same regulatory obligation for assuring qualified workers are performing covered task(s). This is true no matter how many times the work is subcontracted. And, during a state or federal inspection, the operator must be able to produce historical records of qualification for all workers.

19.6 RECORDS AND THEIR RETENTION

Records need to be maintained for active qualifications and previously expired qualifications for 5 years after the individual no longer performs the task. The challenge for most operators is retaining records for the multiple levels of subcontractors that may have been involved. Operators can contract the work but not their regulatory responsibilities. The record retention challenge becomes even more overwhelming if, as part of a special permitting process, the pipeline agreed to maintain records for the life of the facility. This means qualification records for everyone who performed a covered task throughout the lifecycle, from construction to maintenance, must be historically preserved indefinitely.

19.7 TAKING OWNERSHIP WHEN REPUTATION IS ON THE LINE

The first step is to understand the importance of companies taking ownership of their OQ programs to ensure safety and compliance. The discovery of a 2025 violation can result in fines of over $238,809 per day, and for a related series of violations, it's $2,388,090 per day, with higher penalties for violations resulting in severe injury or death. The risk is real to people, property, the environment, company reputation, and their bottom line.

Intuitively, we understand that there is little separation between the operator and the contractor, especially in the public eye. When you hear a pipeline incident being reported, listen carefully to how the individuals involved in the incident are described. Most often, you will hear the workers described as "Insert Pipeline Operator's Name" workers or "Insert Pipeline Operator's Name" contractors. In a few instances, the contract company name will be used, but rarely. The point is, regardless of who is performing the work, the risk and reputation of the pipeline operator is tied directly to the incident. While safety and regulatory requirements seem like equally shared responsibility between operator and contractor, it really is a partnership where the operator must take the lead.

Ensuring OQ program alignment between employees and contractors is a significant challenge for pipeline companies, as inconsistencies in training, evaluation, and compliance standards can lead to safety risks and regulatory gaps. Taking ownership and being engaged in the implementation of your program should include aligning expectations for employees and contractors alike. After all, they are all working on your system and should be held to the same standards. Look for inconsistencies in training, qualification, evaluation, knowledge of your policies and procedures, and recordkeeping. Many of the regulatory gaps can be discovered by comparing the employee qualifying journey to your contractor journey. Can you pass the red face test when you compare the two paths to qualification?

Assessing the current state of your OQ program and how internal employee qualifications and training align with the contractor experience is a solid way to mature and improve your program. Maturing your OQ program is the path to continuous improvement for documenting and measuring program effectiveness. The primary drivers of OQ program maturity are culture, technology, and data.

1. Culture is manifested via a clear commitment from management with sufficient resources to support field personnel in the implementation and spirit of the program. With consistent management support, the expectation to work safely and follow company procedures becomes the norm.
2. Technology allows real-time mobile access to a central repository to enter and access everything, from training records to qualifications. Being able to confirm worker readiness status and enter everything from an evaluation event to OJT is key functionality to avoid delays in getting people on the job site ready to go.
3. Secure data collection and retention are critical to providing evidence of historical compliance and key performance metrics that verify program effectiveness measures.

So, based on the information given, how is your program, and how are you qualifying internal employees and contractors? Is your program maturing? Or are there gaps that you need to address?

Over the years, I have reviewed hundreds of OQ programs to find some are more advanced than others. I have identified five levels of OQ program maturity, starting with ad hoc and growing to developing programs, basic programs, advanced programs, and strategic programs. As I define each category on the continuum of five levels of OQ program maturity, where does your program stand today? And more importantly, what steps can you take to improve your current state?

An *ad hoc* program is the starter set with the following criteria being common:

1. sparse resources;
2. limited awareness of requirements;
3. attention paid to internal employees;
4. not much thought given to how contractors are to be included,
5. disorganized recordkeeping – where necessary data often resides in emails or in someone's desk in the form of hard copies, with no centralized method to capture current or historical qualifications; and
6. no clear links between operations and maintenance procedures and how regulatory work is being accomplished.

A *developing program* is generally a step above the ad hoc arrangement described earlier, with the following features, based on my observations:

1. there is typically an individual, a small part of whose job is to attempt to gather all the requirements and records from a mix of sources;
2. the program is communicated informally throughout the organization;
3. there is usually an attempt to include the contractors, but it is inconsistent;
4. most companies have an O&M in place, though it typically falls short of identifying all the operators' potential covered tasks;
5. the operators rely on whatever records contractors can provide, with no way to validate or trace back how the qualification was put in place;

6. there is no formal task training, and records are captured locally on Excel or Word files; and
7. leveraging data to improve the program is difficult because there are no formal program metrics in place to review.

Then a step above a developing program, there is what I call a *basic program*, often characterized by the following:

1. keeping up with the program is a significant part of someone's job;
2. there is usually an actual dedicated compliance resource to maintain and help implement the program;
3. operators have a formally communicated written OQ program with a task list that matches the O&M work performed on their specific facilities;
4. there is a defined evaluation process;
5. there are scheduled requalification intervals;
6. there is an identified SOC;
7. the abnormal operating conditions have been considered and thought through;
8. this program identifies when training is appropriate;
9. there is a clearly defined management of change;
10. there is a recommendation for appropriate notifications to state and federal agencies;
11. there is a dependable method to capture records and data to measure against standardized performance metrics; and
12. there is also an annual program review that captures basic information to identify opportunities for improvement.

Next, an *advanced program* is even more mature than the basic program and has the following aspects:

1. full-time internal resources dedicated to it, and often additional third-party resources are engaged for support;
2. there is a history of OQ program revisions with documentation of annual reviews to support defined program effectiveness metrics;
3. there is a presence of a task list based on their O&M;
4. there is an established qualification process;
5. there are scheduled requalification intervals;
6. there is an identified SOC;
7. the abnormal operating conditions have been considered and thought through;
8. there is a formalized method to select and authorize evaluators;
9. there is a task-specific training program;
10. there is a way to have instant access to OQ records;
11. there is standardized data capture with historical reporting;
12. there is an active utilization of their annual OQ program review and O&M review with the end goal of comparing them against how work is being performed on the job;

13 there is a process for capturing changes;
14. there is a method to ensure proper notifications to PHMSA, states, management, internal employees, and contractors.

Finally, *strategic program* – the most mature of all programs – is what all the organizations should aim for. Why? Because a strategic program cultivates a safety culture supported by management, uses technology to capture all the required documentation, and uses data to track key performance metrics to continuously improve reporting to make informed decisions. This is usually accomplished through the following elements, though this is not an exhaustive list:

1. there are full-time internal resources dedicated to the administration of the program;
2. there are third-party resources present to support audits of personnel and qualification records;
3. there is an easily producible history of continuous improvement;
4. there is usually a hybrid list of common covered tasks, which are portable, and a set of company-specific tasks that are unique to their operation, tied directly to their O&M;
5. there is a formalized audit process of evaluators, qualified personnel, and records – this is an ongoing QA/QC step that assures compliance;
6. there is an evaluator authorization program, including evaluator training to support the integrity of the qualification process;
7. there is a management of change process to include notifications to PHMSA, states, and all qualified individuals, evaluators, and supervisors;
8. there is a requirement for training for initial and subsequent qualifications, regardless of experience;
9. there is also extensive training on the company's policies and procedures;
10. there are usually other requirements, such as drug and alcohol compliance, site orientation completion, certifications, safety records, insurance, background checks, and positive identification, which are linked to OQ reporting to confirm that the individual meets the ready-to-go-to-work requirements;
11. there is a flexibility that comes with means to provide real-time access to historical records and data, including internal personnel and contractor employees;
12. there is appropriate data retention and purge, according to regulatory retention requirements;
13. there are multiple program effectiveness metrics with trends and improvement history;
14. the annual OQ program review includes actionable steps to correct or enhance the program;
15. the O&M and state-defined requirements help determine and evolve the OQ task list.

19.8 CONCLUSION

Having a basic OQ program in place to meet all the minimum regulatory requirements is a must. There is no ramp-up or grace period for compliance or safety. If you are not there now, it is important to close that gap immediately.

But don't stop with the basics. If you want to save lives, as opposed to meeting the least common denominator, you need to keep advancing your OQ program along the maturity continuum. This will not only allow you to be a leader in qualifying pipeline personnel, but even more importantly, it will equip you with additional tools necessary to reduce the risk of serious incidents and fatalities that can harm people, property, and the environment. Bringing workers home safe from high-hazard jobs makes all the effort worth it.

20 The Current of Change *Challenges and Growth for Utility Providers and Their Contractors*

Simon Goncharenko

20.1 INTRODUCTION

Power generation and transmission have received significant attention, as utility clients and contractors are challenged with both meeting growing demand while also navigating through the increasingly complex set of requirements originating at the global, national, state, and local levels, as well as the corporate commitments. So, this chapter will identify the key risks of the utility industry and the resulting pain points experienced by the contractors, in order to offer some solutions for a sustainably safer tomorrow, which includes satisfying the regulatory requirements, pleasing the consumers, and sending every worker within the industry home in better shape than before. Sounds like a tall order, but it is achievable, so long as the right foundations are in place.

20.2 TODAY'S CHALLENGES IN THE UTILITY INDUSTRY

First things first, what are today's challenges for the utility hiring clients and contractors? The simple answer is that there are many of them, and they are quite complex and multifaceted.

20.3 EXTERNAL CHALLENGES

On the one hand, utility providers today must deal with a long list of external pressures and requirements. Among them are things like:

1. Contending with having to paint within very strict lines established by the global, federal, state, and local regulations.
2. Responding to often unpredictable conditions arising from natural disasters and extreme weather.
3. Facing soaring power consumption amid electric vehicle and data center market growth, just to name a few.
4. Navigating within an aging and often crumbling infrastructure.

DOI: 10.1201/9781003679448-20

5. Meeting increasingly aggressive decarbonization and energy transition goals.
6. Attending to stressed-out consumers who are squeezed by the current economic conditions.
7. Negotiating an often unfriendly media and social media climate.
8. Attracting investments amid uncertainty brought about by the demise of the Chevron doctrine, etc.

On the other hand, there are also internal issues that utility providers and contractors, like any other company today, must negotiate strategically in order to succeed. Among the internal issues are:

1. Poorly trained personnel.
2. Cultural, language, and communication barriers.
3. Lack of proper qualifications.
4. Psychosocial factors, including mental well-being.
5. Intergenerational struggles, etc.

To all of this, one could add the financial pressures of remaining soluble and profitable, without which there is no future. And this list only scratches the surface.

So, the goal of this chapter, then, is to unpack the challenges and identify simple principles, following which will result in sustainably safer tomorrows for the utility clients and contractors and happy and fulfilling jobs for the workers who are in this industry.

20.4 EXTERNAL – REGULATORY CHALLENGES

Take painting within the strict lines of regulatory and corporate requirements, for example. Global organizations and treaties, like the World Economic Forum, United Nations, World Health Organization, Paris Accord, and others, have set out fairly specific requirements and guidelines, which may or may not be enforceable within various geographies around the world. On a national level, the United States serves as a good example of the multifaceted web of regulations for the utility providers. Here, in the United States, utilities are regulated by both the Federal Energy Regulatory Commission and state and municipal public service commissions, both of which exist to ensure that consumers have access to reliable, safe, and secure energy at a reasonable cost. In addition, many of the utility companies, the municipalities that own them, or their holding companies are committed to the Environmental Social Governance requirements, which can consume significant time, energy, and resources. This often results in equal or similar expectations being placed on their contractors, who often find themselves without much choice in the matter.

20.5 EXTERNAL – WEATHER CHALLENGES

Weather has always been unpredictable and even the best of planning and preparations can still be quickly exhausted by either a significant storm, like Hurricanes

Andrew or Katrina, Typhoon Tip or Haiyan, or a series of smaller events happening almost in tandem with one another. This was witnessed in Houston in 2024, when Beryl, which struck the Texas coast as merely a Category 1 hurricane on the morning of July 8 ended up causing the death of at least 36 people and leaving almost three million residents without power for close to 2 weeks amid the sweltering summer heat. The reason for this unprecedented devastation by a relatively minor storm was that it was the third storm to strike the nation's fourth largest city in less than 2 months. Since repair crews often perform the quickest possible fixes to restore power, after which they work on the longer-term, more permanent fixes, in this case, there was no time for them to complete the permanent fixes, which left Houston area exposed to significant impacts.

20.6 EXTERNAL – INCREASED DEMAND CHALLENGES

The world's growth in demand for electricity, for example, in 2024 and 2025 and beyond is forecast to be among the highest levels in the past two decades, according to the International Energy Agency. Two of the biggest drivers for this load growth are data centers and electric vehicles, though the demand for the latter seems falling off a bit at the present.

The continued expansion of the Internet of Things and AI is fueling a tremendous uptick in the data center development and construction. According to the American Public Power Association, the combined expansion of traditional and AI data centers, along with chip foundries, will increase demand cumulatively by 177 terawatt-hours from 2023 to 2030, reaching a total of 307 TWh.

Electric vehicles were expected to emerge as the second major catalyst for power demand growth in the United States and around the world up to 2030, though at the present it is unclear if that prediction will be materialized. As an example, in 2023, electricity consumption in the United States transportation sector totaled 18.3 TWh. By 2030, this figure is projected to soar to 131 TWh. Between data centers and EVs, US electricity demand is projected to increase by nearly 300 TWh by 2030.

Likewise, the end-use electricity demand in Canada is growing by as much as 35% between 2005 and 2050. To meet this tremendous increase in demand, all of Western Europe, with the exception of Germany, is accelerating nuclear power programs.

20.7 EXTERNAL – INFRASTRUCTURE CHALLENGES

The growth in demand is especially challenging, given the present state of the utility infrastructure. Experts say that US infrastructure is both dangerously overstretched and lagging behind that of its economic competitors, particularly China. While there has been a significant amount of upgrade projects completed in the past decade in certain parts of the United States, there is still much work to be done, as the average age of the transmission infrastructure in the United States is 40 years and, in the United Kingdom, dating all the way back to World War 2. The estimated cost of the transmission modernization in the United States ranges from $1.5 to $2 trillion by 2030, and this is on the low end.

20.8 EXTERNAL – DECARBONIZATION CHALLENGES

Then there is the push for energy transition. A growing group of US electric companies has committed to reducing carbon emissions by 80% by 2030. In fact, at present, 23 US states, as well as the District of Columbia and Puerto Rico, have adopted some form of 100% clean energy goals. Yet at the same time, most stakeholders are underprepared for the enormous investment that meeting these decarbonization goals will require.

Meeting decarbonization goals means the United States needs anywhere from 1.3 to 2.9 times current transmission capacity, according to the National Renewable Energy Laboratory estimate. How so? Here is an example that perfectly illustrates this dilemma. Nearly 60% of New York State's electricity demand comes from New York City. This demand was, until recently, largely met by fossil-fuel plants in and around the city. As those are replaced by hydroelectric plants, wind farms, and solar plants, the power supply moves further away from the source of demand, putting more pressure on the transmission system to shuttle volumes of electricity it was never built to handle, thus the reason for increasing transmission capacity. And the United States is not alone in the need to upgrade its infrastructure for the sake of energy transition. Germany had to upgrade its capital spend predictions to €61bn between now and 2030, which is almost double its initial estimate of €33bn from 2017.

20.9 EXTERNAL – CHALLENGE OF ATTENDING TO CONSUMERS SQUEEZED BY ECONOMY

The state of the economy and price increases on everything – from groceries and gasoline to housing and insurance – are making it more difficult for the average consumer to afford something as basic as their utility bills. This presents a challenge for the providers in balancing their own profit and loss statements while finding creative ways to help their customers stay afloat. This challenge is further exacerbated by the media and social media's often unfair coverage of the utility providers, who are frequently pitted against the consumers and painted in a negative light. And now, the recent Supreme Court Chevron decision is also going to potentially impact investor willingness to participate in the electrical transmission industry.

20.10 INTERNAL CHALLENGES

While this chapter has gone to great lengths to establish the extensive nature of the external pressures and challenges faced by the utility providers, some may wonder why this was necessary and how it is relevant to the topic at hand. The fact of the matter is that a purely cursory overview of the external challenges may result in a lack of appreciation for the tremendous amount of pressure that these challenges add to the internal issues. Because at the end of the day, those individuals who comprise the workforce within the industry are the very same frazzled people who are impacted by the weight of the external challenges.

20.11 INTERNAL – WORKFORCE TRAINING ISSUES

A significant variation in the training standards from state to state and from company to company means that not all utility workers are equally trained. Some are not willing to put in the 3–4 years of fieldwork that it takes to become a fully qualified lineman able to work directly on electrical equipment. And others may need to be retrained from the previously wrong ways in which they had been brought up.

20.12 INTERNAL – CULTURAL BARRIERS

Wars, conflicts, and migration influxes have contributed to the increasingly diverse workforce around the world. Taking the United States as an example, with a recent significant influx of people from all over the world into the country, language and communication barriers are becoming more of an issue on a daily basis. Greater globalization of the workforce means that interpersonal challenges are much deeper than just language and communication. Cultural values and norms must be considered, such as what constitutes right and wrong or good work versus "good enough" work, among other things. Having grown up outside of the United States, I personally experienced many of the things mentioned earlier. These challenges necessitate consistent efforts on the part of the employer to ensure clarity.

20.13 INTERNAL – EMPLOYEE QUALIFICATION STANDARDS

Fitness, drug and alcohol, education, certifications, and other employee qualifications must be met and managed to enter the utility industry ranks. Many of these qualifications must be renewed and maintained annually. When access to high-risk environments must be controlled based on these qualifications – and managed not only for employees but also for contractors and subcontractors – keeping up with all of this could quickly and easily become unmanageable. The good news is that there are third-party providers like Veriforce, whose whole job is to qualify, verify, and track these data points for their clients.

20.14 INTERNAL – PSYCHOSOCIAL ISSUES

Various psychosocial challenges can limit and greatly impair utility workers' ability to focus and perform their work effectively and safely. Among these are anxiety and depression, eating disorders, and psychotic disorders. The psychotic disorders are particularly interesting as many of them are caused by cannabis use, which is rising significantly today, both in the United States and many other countries around the globe. Astute employers will ensure the presence of appropriate policies and protocols to control the usage of both drugs and alcohol within their jurisdictions and offer assistance options to those who need them.

20.15 INTERNAL – INTERGENERATIONAL CHALLENGES

We are experiencing the first occurrence in the work history when five different generations may be found in the workplace at the same time, all the way from the

Silent Generation to the Gen Z. This is especially critical as each generation has a distinctly different set of values, communication styles, approach to work, and much more. If tackled with the right tools, the intergenerational aspect can be a significant asset for an organization, but it can also present tremendous challenges if it is viewed as a tug-of-war.

20.16 PRACTICAL SOLUTIONS FOR THE UTILITIES PROVIDERS AND CONTRACTORS

What steps must be taken to overcome the challenges and construct the foundation for a sustainably safer tomorrow in utilities?

On the governmental level, be it national governments, state or regional legislators, or municipalities, care must be exercised so as not to produce conflicting regulations and unintended consequences, much like in the case of the loosening of the marijuana restrictions while continuing to test for its presence.

On the corporate level, there are no shortcuts to achieving the right culture. A two-pronged approach should be considered: Tackling the culture and obtaining the right tools. The culture of safer thinking begins with the executive office and must be integrated into every level of your organization, from communication to company functions, to values, to mission, to expenditures, to rewards and recognition, to planning and scheduling, and everything else in between. The culture of safer work is arrived at by creating systems that make it difficult for incidents to occur, make it impossible for those incidents to kill or hurt someone, and make it easy for everyone to do their daily jobs safely.

While building a safer culture may take time and consistent and considerable effort, the benefits of doing so cannot be overestimated, ranging from employee satisfaction, higher engagement, increased productivity, decreased absenteeism, increased profits, and happier clients. In short, as someone once said, you can buy your employees' hands and backs, but you must win their hearts and minds. The right culture does that.

Obtaining the right tools is equally important. There is nothing more frustrating and demoralizing than your personnel wanting to do the right thing because of the progress you are making in culture building, yet unable to do it because of the lack of tools. Of the many tools to be considered, how you manage your supply chain risk must not be overlooked. Find the right solution, and you will find peace of mind related to operator qualifications, tracking and administering drug and alcohol requirements, contractor management, contractor communication, staying on top of training and qualifications, insurance agreements, etc. You will additionally gain immediate relief in the areas of risk management and mitigation, streamlined communication, documentation and compliance tracking, emergency preparedness, and data-driven decision-making.

On the individual utility worker level, your employees need to remember that they have one body. What you do to it at home will most certainly impact your ability to perform effectively, efficiently, and safely at work. The opposite is also true.

So, how should your employees treat their bodies in order to ensure they come home safe from their workplaces, which happen to be high-hazard in the utility industry? Pay attention to things like your stress level, your work–life balance, your nutrition, your alcohol intake, your choice to partake or not to partake in recreational drugs, your mental state, your spiritual state, and your sleep habits. Though there is often a temptation to feel invincible, our bodies will only respond in accordance with how we treat them. This is true with any other tool or equipment you use in your daily activities, and it is true of your body as well.

Take sleep, for example. Is your personal average number of sleep hours within the recommended range for your age (for adults 18–64 years, it should be seven–nine hours)? If not, you place yourself in a higher-risk category for developing several medical conditions, including inflammation, stroke, heart attacks, Parkinson's, Alzheimer's, and anxiety and depression. Additionally, your productivity is going to decrease, as will your ability to focus, concentrate, and maintain your emotions under control.

Then you also place yourself and those working with you at a higher risk for workplace injuries and death. This chapter's scope does not allow an in-depth discussion of all the significant incidents that resulted from sleep deprivation, like the Three Mile Island Nuclear Disaster, Space Shuttle Challenger explosion, Chernobyl Nuclear Plant devastation, or Exxon Valdez spill. Suffice it to say that the list of incidents is long and quite convincing. While no one in their right mind would want to add their employer's name to the list, many, if not most, of us knowingly take a lot of steps that all but bring us to that edge on almost a daily basis.

20.17 SUMMARY AND KEY TAKEAWAYS

In conclusion, on the one hand, utilities, unlike many other industries, are facing some very unique and complex challenges today. These cannot be ignored, for they will not go away in this manner. These challenges are comprised of external pressures from outside utility providers, such as regulatory environments, infrastructure conditions, and decarbonization goals. The challenges faced by the utility industry must be tackled head-on in a wise and methodical manner.

At the same time, on the other hand, some of the internal personnel and culture issues that are faced by the utility providers and their contractors today are really no different from any other industry. Every employer must overcome problems with sufficient training, clear communication, and intergenerational challenges. Those who are fervently planting the seeds of building the right culture and investing in their personnel via the right contractor management tools will reap the benefits of a sustainably safer tomorrow.

21 The Role of Effective Assurance While Constructing the International Space Station

Jim Wetherbee

21.1 INTRODUCTION

In the mid-1990s, US President Bill Clinton directed NASA Administrator Dan Goldin to review the progress of Space Station *Freedom*, the first US space station being designed since the *Skylab* program in the 1970s. The *Freedom* program was headed for trouble. Every aspect of its design – power, life support, weight, launch performance, cost – was in the red and was predicted to fail. Based on the review, the *Freedom* program was disbanded, and NASA relocated the program to the Johnson Space Center in Houston, TX. The Director of the Center, George Abbey, was simultaneously running human spaceflight for the agency. Under the leadership of Col. Randy Brinkley, USMC (Ret.), the new space station design was called the *International Space Station* (ISS).

Director Abbey knew very well the immense challenges of leading an effort to design, fabricate, integrate, launch, and construct in orbit a technological marvel, such as the ISS. He learned how daunting the task would be as he lived through the *Apollo* program tragedy when three astronauts were killed in a launch pad fire on January 27, 1967. After the fire, Dr. George Low was called in to rescue the *Apollo* program from the ashes. Dr. Low was responsible for completing President Kennedy's challenge of "achieving the goal, before this decade is out, of landing a man on the moon . . . ." The astronauts especially appreciated the next part, ". . . and returning him safely to the earth."

Dr. Low recognized the monumental significance of leading hundreds of thousands of personnel, including thousands of contractors, in a national effort to leave the gravitational bonds of Earth for the first time in history and fly to the moon. At the time, Dr. Low's assistant was George Abbey, the young, brilliant engineer and Air Force officer. Under the mentorship of Dr. Low, Abbey learned the lessons that would serve our nation well decades later as NASA was again trying to rescue human

DOI: 10.1201/9781003679448-21

spaceflight from the tragedy of the *Challenger* disaster and simultaneously design a new space station in the 1990s.

No single organization can conduct all the operations required to accomplish complicated objectives, such as constructing the *International Space Station* in low Earth orbit. Many organizations and groups must contribute to the mission simultaneously. Making strategic decisions is most challenging for the executives, who must be able to predict future progress toward mission goals with multiple organizations. Unfortunately, no single person can understand how well each entity is performing in service of the mission (Wetherbee, *Controlling Risk*).

Dr. Low and George Abbey rescued the *Apollo* Program in 1967, and now as Director, Abbey led the effort in the 1990s with the ISS Program. His secret? An especially insightful and powerful process. To succeed – and prevent disasters – the executives relied on an effective assurance process to help them predict the performance of each entity.

In an operational organization, providing assurance means a person is giving confidence about future performance to another person or group, based on observations or assessments of past and current performance. An assurance team observes a sample of operations and evaluates the quality of performance. From their assessments, the assurance team leader uses judgment based on extensive experience to give an executive (or governing board) a prediction, with levels of confidence, that the assessed entity will continue to deliver desired performance in the future – or won't continue. That confidence should be qualified with derived and estimated uncertainties, based on how detailed and complete the observations or inspections were when assessing the quality of past and current performance (Wetherbee, *Controlling Risk*).

Few modern companies seem to have mastered the daunting task of leading contractors and subcontractors undertaking complex endeavors to ensure reliable outcomes. Many managers in various organizations do not seem to understand how to create an effective process of assurance, nor do they seem to understand the value of assurance. Executives and managers could avoid major accidents if they understood how well – or how poorly – supporting organizations were performing if they implemented an effective assurance process. Without this, the executives might have erroneous confidence that organizations would continue to perform well.

Assurance is a *prediction of future performance*, based on an *evaluation* of the *quality* of *past and current performance*, including levels of confidence and uncertainty.

To make strategic decisions, executives must predict future progress toward mission goals across multiple organizations. To succeed – and prevent disaster – executives must rely on an effective assurance process to help them predict the performance of each entity.

An assurance team observes a sample of operations and evaluates the quality of performance. From their assessment, based on their professional experience, the assurance team leader can give an executive (or governing board) *confidence that* the assessed entity will *continue to deliver desired performance* in the future. Conversely, an experienced assurance team can pick up signals and signs that indicate that the assessed entity is less likely to continue to deliver desired performance in the future. That confidence should be qualified with derived and estimated *uncertainties*, based on how detailed and complete the observations or inspections were when assessing the quality of past and current performance.

In summary, providing assurance involves the following three steps. First, gathering initial data via sampling, observations, testing, audits, and verification of past and current performance of an entity. Second, analyzing the data in order to understand the story that it is revealing. The data will be assessed, analyzed, evaluated, and trended in this step, with the goal of not only understanding more fully the present reality but also predicting the future states. Finally, the third step involves interpreting and applying the information to the present reality to accurately forecast future performance, especially related to the contractors' abilities to meet the schedule deadlines, quality, cost, and other deliverables.

Many organizations diligently conduct the processes of sampling, observations, testing, audits, and verification, specified in the list given earlier, when managing contractor companies. Few organizations seem to conduct the subsequent process of using the vast experience and superior judgment of an assurance team, led by a sage leader, who takes the information collected and makes a valuable prediction of future performance. With this prediction, the managing executives can make the best decisions in the pursuit of operating excellence with the various contractors and subcontractor entities that are supporting the organization's mission. This final prediction is the essence of the assurance process.

It was an effective assurance process, implemented and adhered to by George Abbey, that enabled him to succeed in leading the team that created the inspiring *International Space Station*. This shining beacon has been passing overhead for 25 years, illustrating how multiple nations and thousands of contractors can come together to create a technological masterpiece, while inspiring generations of doctors, scientists, engineers, and leaders into the future.

FURTHER READING

Wetherbee, Jim. (2019). *Controlling Risk – in a Dangerous World.* New York, NY: Morgan James.

22 The Impact of Chevron Deference on American Clients and Contractors

Simon Goncharenko

22.1 INTRODUCTION

In this short chapter, we will review the decision to strike the Chevron Doctrine, examine Chief Justice Roberts' clarification, and consider the broader implications for the clients, contractors, and their respective employees.

22.2 WHAT IS THE CHEVRON DOCTRINE?

It all stemmed from a narrow question of fishery management law: whether the National Marine Fisheries Service's (NMFS) requirement that certain vessels pay for federal observers onboard their boats was consistent with the Magnuson-Stevens Act. But the root issue of the case was whether the court administering the case should defer to a federal agency's interpretation or whether it should decide it on its own (Chevron USA Inc. v. Natural Resources Defense Council, Inc., 467 US 837 (1984)). In that case, the court deferred to NMFS in its decision. This became known as the Chevron doctrine, and this methodology has been used in adjudicating more than 18,000 court cases and arriving at more than 3,000 administrative decisions for the past 40 years. In courts, the judicial branch deferred to "permissible" agency interpretations of the statutes those agencies administer – even when a reviewing court read the statute differently.

What ended up happening then was that each case was resolved in favor of the government's challenges. So, the bottom line was that if an employer, be they a hiring client or a contractor, wanted to challenge the government, the government generally won under the Chevron doctrine.

22.3 WHAT DOES THE JUNE 28, 2024 DECISION DO TO THE PREVIOUS 18,000 CASES CITING CHEVRON?

Chief Justice Roberts indicated that the court's decision would not require earlier cases that relied on Chevron to be overturned. Roberts wrote,

> Mere reliance on Chevron cannot constitute a "special justification" for overruling a decision upholding agency action, because to say a precedent relied on Chevron is, at

DOI: 10.1201/9781003679448-22

best, just an argument that the precedent was wrongly decided – which is not enough, standing alone, to overrule the case.

22.4 WILL OVERTURN HELP OR HURT CLIENT AND CONTRACTOR EMPLOYERS AND EMPLOYEES?

For those who believe that the government will always have the employer's or employees' best interests in mind, the recent Supreme Court decision is damaging. It may mean lengthier rulemaking processes, more legal challenges from the employers, and less ability for the federal agencies to dictate the executive branch's dogma or each new administration's philosophy for all.

One of the underlying ideas behind the camp that believes this Supreme Court's decision will hurt the employees is the conviction that the average client and contractor employer is intent on mistreating and abusing their employees to the extent allowable by law. And there is no doubt that some employers may indeed act that way. But most do not.

Why? Because employees always have the option to vote with their feet, leaving truly terrible client and contractor employers the moment they have a chance for a better job. Common sense suggests that burdensome, stringent, or arbitrary interpretations by various agencies are unnecessary to rein in bad employers, as their abusive practices typically lead to their own downfall.

Additionally, those employers that will choose to litigate against an agency will still have to establish principally that an agency action was arbitrary and capricious, abused discretion, or a violation of the law to prevail on a challenge to an agency adjudication or rulemaking. In other words, courts will no longer defer to an agency's interpretation of a statute, but those challenging it will still have a relatively high burden of proof.

However, for those who believe in limited government, the overturn of the Chevron deference is a restoration of the balance between the three separate and independent branches of government: legislative, executive, and judicial. The executive branch of the government, and federal agencies in particular, will no longer be able to utilize arbitrary or capricious methods in interpreting the law. This limitation of the federal government, according to this view, can only help employers and employees alike.

Those in this camp simply cite some recent examples of the federal agencies' arbitrary overreach, which resulted in hurting instead of helping both employers and employees. One such case is the Securities and Exchange Commission's Environmental, Social and Governance (ESG) rule requiring registrants to disclose information that can and has been widely politicized and weaponized against employers, resulting in lost contracts, difficulties obtaining loans and lines of credit, reputational damages, etc. This rule was actively resisted by a significant number of American industries and employers due to its burdensome requirements overtaxing businesses, some of whom are still struggling to recover from COVID-19.

Another example of an agency overreach was OSHA's 2021 COVID-19 Vaccine Mandate, based on poorly researched data, which intended to force up to 80 million

American workers to choose between their jobs and experimental shots. This emergency standard had to be blocked by the Supreme Court.

All in all, while it may still be too early to know the full extent of the Supreme Court's Chevron ruling, this chapter established that its effect upon the US employers and employees alike largely depends on one's perspective regarding the role of the federal government in our everyday lives.

SOURCES CITED

Goncharenko, Simon. (2024). *Will US Supreme Court's Overturn of Chevron Deference Help Or Hurt American Employers and Employees?* July 11. www.linkedin.com/pulse/us-supreme-courts-overturn-chevron-deference-help-simon-hjtnc/

Weiss, Eric P. (2024). *SCOTUS Rules That Courts No Longer Have to Give Deference to Interpretations of Federal Regulations.* July 14. https://scalirasmussen.com/news/2024/07/14/scotus-rules-that-courts-no-longer-have-to-give-deference-to-interpretations-of#:~:text=%E2%80%9CMere%20reliance%20on%20Chevron%20cannot,to%20comply%20with%20the%20law.

Index

For Product Safety Concerns and Information please contact our
EU representative GPSR@taylorandfrancis.com Taylor & Francis
Verlag GmbH, Kaufingerstraße 24, 80331 München, Germany